BISCUIT, COOKIE AND CRACKER PRODUCTION

BISCUIT, COOKIE AND CRACKER PRODUCTION

Process, Production and Packaging Equipment

IAIN DAVIDSON
Director, Baker Pacific Ltd.

ELSEVIER

ACADEMIC PRESS
An imprint of Elsevier

Academic Press is an imprint of Elsevier
125 London Wall, London EC2Y 5AS, United Kingdom
525 B Street, Suite 1650, San Diego, CA 92101, United States
50 Hampshire Street, 5th Floor, Cambridge, MA 02139, United States
The Boulevard, Langford Lane, Kidlington, Oxford OX5 1GB, United Kingdom

Library of Congress Cataloging-in-Publication Data
A catalog record for this book is available from the Library of Congress

British Library Cataloguing-in-Publication Data
A catalogue record for this book is available from the British Library

ISBN: 978-0-12-815579-0

For information on all Academic Press publications
visit our website at https://www.elsevier.com/books-and-journals

Publisher: Andre Gerhard Wolff
Acquisition Editor: Nina Rosa de Araujo Bandeira
Editorial Project Manager: Hilary Carr
Production Project Manager: Denny Mansingh
Cover Designer: Victoria Pearson

Typeset by SPi Global, India

CONTENTS

BIOGRAPHY

Iain Davidson, Director Baker Pacific Ltd., graduated from the school of industrial design (engineering) at Royal College of Art in London in 1965 and joined Baker Perkins Ltd. He was an industrial design engineer, working in the technical department on the design of new biscuit, bakery and candy processing machines until 1975, gaining a thorough technical knowledge of the machines and processes.

In 1975, he was appointed market development manager at Baker Perkins, involved in developing the Baker Perkins forward planning for new business, product development and acquisitions. In 1979, he became international sales manager with responsibility for the biscuit business in Asia and Africa.

In 1990, he was appointed regional manager of Asia Pacific for Baker Perkins and relocated to Indonesia and later in 1997 to China. His appointments included managing director of Baker Perkins (Hong Kong) Ltd. and director of Baker Perkins Japan KK.

In 1990, he negotiated an agreement with the Liaoning Foreign Trade Corporation in Dalian to establish a manufacturing facility for biscuit ovens. This was successful in manufacturing ovens under the supervision of Baker Perkins Japan KK engineers.

BAKER PACIFIC LTD

In 2000, he left Baker Perkins Ltd. and established his own company in Indonesia, PT Baker Pacific Mandiri. The company provided consultancy in the biscuit and confectionery industries in Asia. As the business outside Indonesia grew, Baker Pacific Ltd. was established in Hong Kong in 2004 and is now our principal operating company, providing process technology and machinery for the biscuit, chocolate and candy industries. Baker Pacific Ltd. established a manufacturing facility for biscuit ovens and manufactured ovens in China, India and Indonesia.

During recent years, he has provided training programmes for technical staff in biscuit bakeries, both courses presented on site and also provision of presentation materials for in-house programmes. He has completed a technical manual *Biscuit Baking Technology, Second Edition* published in 2016 by Academic Press, an imprint of Elsevier.

EXPERIENCE IN THE BISCUIT INDUSTRY

- Engineering design of biscuit process machines including a range of baking ovens
- Biscuit baking oven manufacture in China, Indonesia and India
- Sales and marketing of biscuit production equipment in Europe, Asia, North America and Africa
- Project management and service for biscuit production lines
- Preparation and presentation of training programmes for technical staff

Acknowledgements

I wish to record the contributions to our business of Jean-Marc Simandoux, John Lilley, Steve Eldridge, David Young, Sunny Xu and the late Glyn Sykes. I am also indebted to Michael Greaves for his advice on chocolate and on quality control and to all our colleagues over the years in the baking industry and at Baker Perkins Ltd.

In addition, I would like to thank all the companies who have contributed valuable information and photos of the equipment and process for *Biscuit, Cookie and Cracker Production.*

INTRODUCTION

Biscuits are the first and best convenience food:

- Require no preparation—eaten straight from the pack
- Have a long shelf life, 6–12 months
- Nutritious, providing protein and energy
- Can carry vital minerals and vitamins

Biscuits are a truly worldwide food. They are manufactured in almost every country with similar equipment and process.

Our book aims to provide students of food technology and engineering everywhere with an understanding of the complete production system from the raw ingredients to the final packaged products.

It will be of value to people working in the biscuit industry to give a fundamental understanding of the production in an easily understandable form with many illustrations. The book is an ideal introduction to people joining a biscuit manufacturer in almost any role.

The format of the book is based on the maxim *if I read it I forget it, if I see it I remember it.* The book has over 200 illustrations, photos of the machines and drawings, which show the complete process from the delivery of the ingredients to the bakery to the final packaging of the biscuits.

Our book is a way of sharing over 50 years of experience in the biscuit baking industry worldwide. As engineers, we have worked with first-class engineers, food technologists, bakers and confectioners in Europe, North and South America, Africa, Asia and Australasia. Our customers have included the largest multinationals and many local businesses.

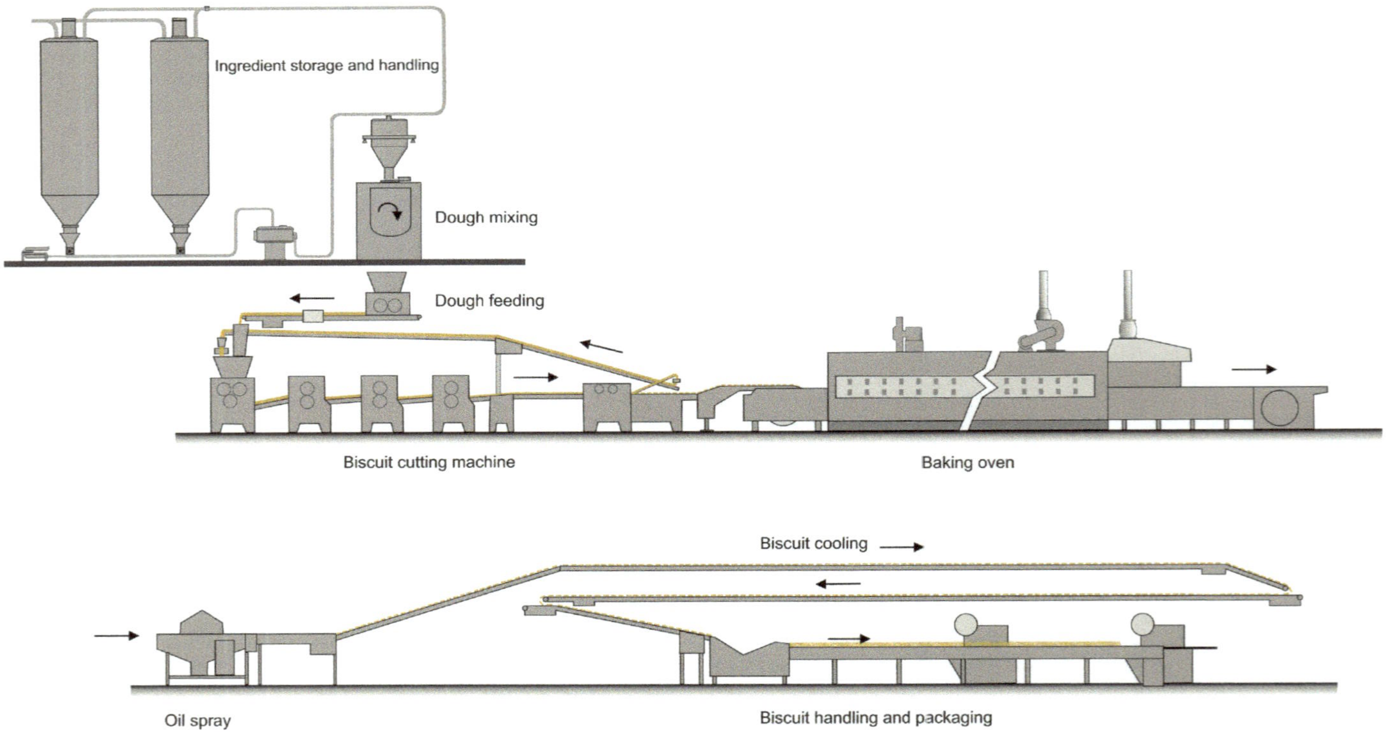

OVERVIEW OF A BISCUIT PRODUCTION LINE

CHAPTER 1

The Biscuits

1.1 BISCUITS ARE THE FIRST AND BEST CONVENIENCE FOOD

The name 'biscuit' derives from the Latin (bis coctus) and later the French (bescuit) meaning twice cooked. The name refers to products, generally made from wheat flour, that were baked and then dried in a slow oven. Although the term 'biscuits' is widely used in many countries, in the United States, the products are called 'cookies and crackers'.

Biscuits have a long history, different forms of 'twice baked' food being made by the Romans, Persians and Egyptians in very early times. They were consumed by sailors and armies and by poor people as a cheap form of sustenance. Biscuits became a staple food for European sailors voyaging to Asia, Africa, Australasia and the Americas.

Biscuits have a long shelf life; are eaten straight from the pack; and are nutritious and available in many forms, both sweet and savoury. Biscuits have many functional forms, enriched with calcium, iron and vitamins and formulated for infants, children and the elderly and for those with special needs such as gluten-free foods.

Biscuits are eaten throughout the world, and many types are familiar in different countries, for example, soda crackers, snack crackers, hard sweet biscuits, short biscuits and cookies. This contrasts with bread, where each country has a very particular type of bread.

There are also many special products popular in their country of origin:

England: Digestive, cream cracker and Rich Tea

France: Petit Beurre

Germany: Lebkuchen

India: Glucose

Italy: Breakfast and Frollini biscuits

Korea: Choco pie

The Netherlands: Speculaas

Scotland: Shortbread

The United States: Soda crackers, saltines and chocolate chip cookies

Biscuit, Cookie and Cracker Production
https://doi.org/10.1016/B978-0-12-815579-0.00001-5

Biscuit manufacturing processes, production and packaging machines are very similar in all countries, and there are now several major international suppliers of production equipment:

- Baker Perkins Ltd., the United Kingdom and the United States
- Dingson Food Machinery Ltd., China
- GEA Group
- Haas Group, Austria (the group now includes Meincke from Denmark and De Vuurslag from the Netherlands)
- Ima Forni, Italy
- Laser S.r.l., Italy
- Middleby Corporation, the United States
- Reading Bakery Systems, Inc., the United States

1.2 CATEGORIES OF BISCUITS

Biscuits broadly fall into four categories, distinguished by their recipes and process: crackers, hard sweet and semi-sweet biscuits, short-dough biscuits and cookies (including filled cookies). Each category and each product type require a particular mixing, forming and baking process.

An example is given of the formulation and process for a product in each main biscuit category. Note that all formulations, recipes and process information given are guides only and their use depends on the local ingredients and production equipment. In each case, the recipes and process will require development during commissioning to suit the particular ingredients and equipment available.

1.3 BISCUIT MAKING PROCESS

In general, the biscuit making process follows the main steps shown below. Mixing and fermentation or dough standing time is usually a batch process. The forming, baking, oil spraying and cooling are continuous operations with a high degree of automation. Packing is generally off-line, unless the line is dedicated to a single product.

1.3.1 Crackers

A wide range of products are characterised by crispy, open texture and savoury flavours.

Crackers include soda and saltine crackers, cream crackers, snack crackers, water biscuits, puff biscuits, 'Maltkist' (sugar-topped crackers), 'TUC' type, 'Ritz' type, vegetable crackers and calcium crackers (Fig. 1.1).

Fig. 1.1 Crackers.

In general, crackers may have some of the following features that influence the baking process:

- Doughs are leavened and fermented with ingredients such as yeast, ammonium bicarbonate and sodium bicarbonate.
- Doughs generally have a high water content (15%–25%).

- Cracker doughs are often laminated (the dough sheet is made up from multiple thin layers).
- Cracker doughs spring or lift in the first part of the oven to achieve an open, flaky texture.
  This process requires humidity and high heat input (Fig. 1.2).

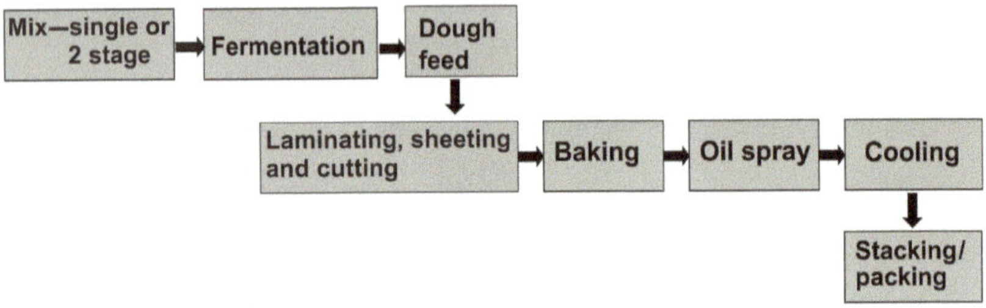

Fig. 1.2 Typical process flow diagram for crackers. Mixing and fermentation are usually batch processes; forming, baking, oil spraying and cooling are continuous in-line processes.

- Some crackers are cut and baked in strips or complete sheets and broken into individual biscuits after baking.
- Some crackers require a colour contrast between dark blisters and a pale background colour.
- Traditional English crackers such as cream crackers and water biscuits are normally baked on light wire-mesh bands.
- Traditional American crackers, such as soda or saltine, are baked on heavy mesh (compound balanced weave) oven bands. These bands are preheated to transfer heat rapidly by conduction into the dough pieces.
- Crackers are baked to low moisture contents (1.5%–2.5%), which requires a high energy input.

1.3.2 Semi-Sweet Biscuits

Examples of semi-sweet biscuits are Marie, Petit Beurre, Rich Tea, Arrowroot, and breakfast biscuits. They are characterised by an even, attractive colour and texture and good volume (Fig. 1.3).

Doughs for semi-sweet biscuits have the following features (Fig. 1.4):

- Doughs have strong, developed gluten that gives an elastic dough, which is sheeted and cut. It often shrinks in the first stage of baking.
- Doughs have low sugar and fat.
- Doughs have water contents typically of around 12%.

Fig. 1.3 Semi-sweet biscuits.

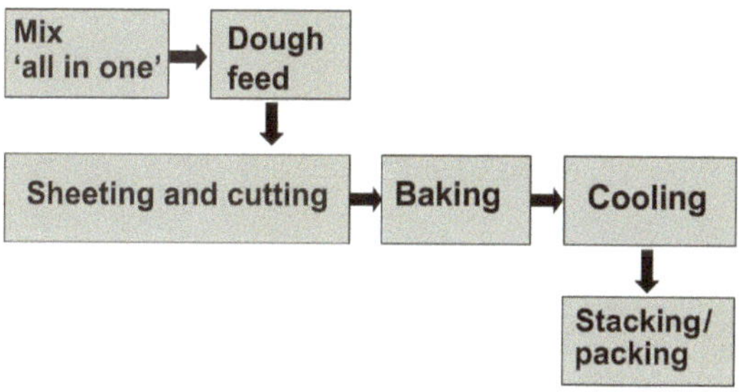

Fig. 1.4 Process flow diagram for typical semi-sweet biscuits.

- Biscuits are normally baked on a wire-mesh band (except for Marie, which is traditionally baked on a steel band).
- Humidity in the first part of the baking is important to achieve good volume and a smooth surface sheen.
- Biscuits are baked to low moisture contents, around 1.5%–3.0%.

1.3.3 Short Doughs: Rotary Moulded Biscuits

This is another wide biscuit category with many designs. The doughs are short with higher fat and sugar contents than the crackers and semi-sweet biscuits. This is the simplest category for dough piece forming process (rotary moulding), and so, these products are very widely produced, often in very high volumes. The following list of typical products would be very extensive and contain many local or regional specialities: malted milk, glucose, Lincoln, digestive, nice, shortbread, custard cream from Britain, Italian frollini, Dutch speculaas, caramelised biscuits and glucose from India.

Short-dough biscuits are often fortified with vitamins and minerals and provide a nutritious convenience food. Short-dough biscuits are also frequently used for sandwich products (Figs 1.5 and 1.6).

- Doughs have a low water content but more fat and sugar than the semi-sweet biscuits.
- High humidity in the first part of the baking process allows the biscuit structure to form.
- Relatively slow baking at comparatively low temperatures is needed.

1.3.4 Cookies

A very wide and popular category includes chocolate chip cookies, butter cookies, two-dough cookies, centre filled cookies, fig bars, fruit bars, extruded cookies and cookies with many types of inclusion such as nuts, raisins, coconut and chocolate chips (Figs 1.7–1.9).

Cookies have the following process characteristics (Fig. 1.10):

- Very soft doughs are deposited directly onto the oven band.
- Recipes have high fat and sugar.
- Long baking times with relatively low baking temperatures are needed.
- All products are baked on steel bands.
- High humidity is required in the first oven zones to allow the dough to spread on the oven band.

Fig. 1.5 Short-dough biscuits.

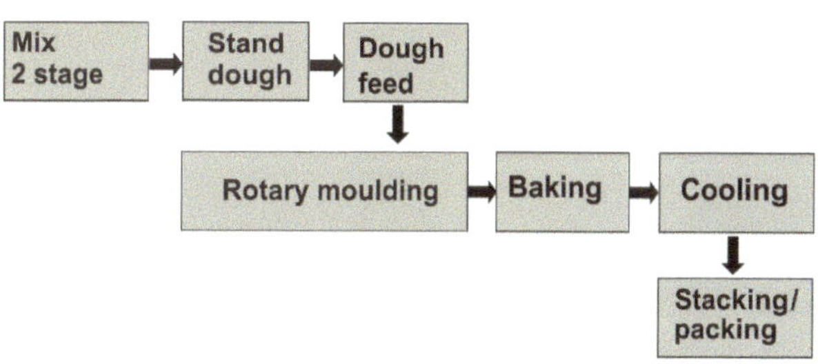

Fig. 1.6 Process flow diagram for typical short-dough biscuits.

Fig. 1.7 Cookies.

Fig. 1.8 Filled cookies.

Fig. 1.9 Danish butter cookies that include deposited cookies and short-dough biscuits.

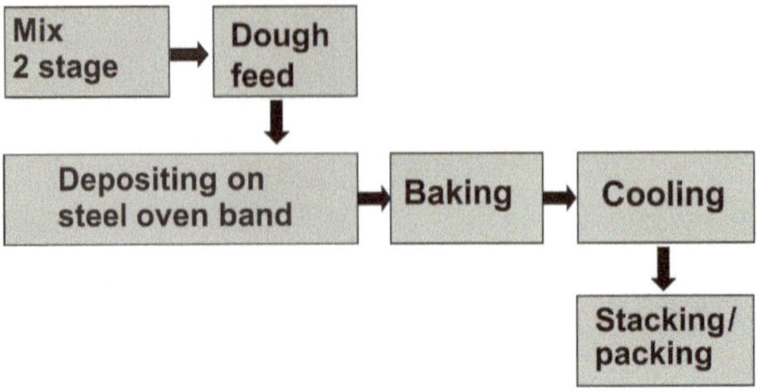

Fig. 1.10 Process flow diagram for cookies.

1.3.5 Long Shelf-Life Cakes, Snack Cakes

A variety of snack cakes are produced on continuous tunnel ovens. These include layer cakes, baked in a complete continuous sheet, which is subsequently cooled, slit and cut into individual products; snack cakes baked individually in tins or pans; and Korean 'pies' that are soft deposited doughs, baked on a steel band (Figs 1.11–1.13).

- Cakes are produced from soft batters with relatively low viscosity.
- Some cakes (layer cakes) are baked in complete sheets of batter deposited on steel bands.
- Some snack cakes are baked in pans that are carried through the oven on chain tracks.
- Korean 'pies' are baked on steel bands as deposited cookies.

Fig. 1.11 Korean pies.

Fig. 1.12 Snack cakes.

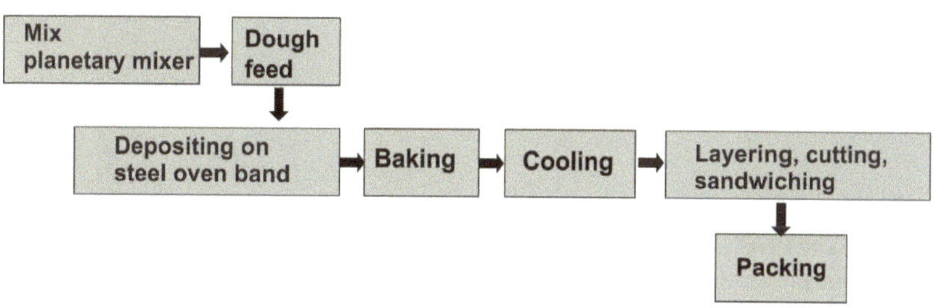

Fig. 1.13 Process flow diagram for cakes.

1.4 SUMMARY

We have listed the main product types:

- Crackers
- Semi-sweet biscuits
- Soft-dough biscuits
- Cookies
- Snack cakes

It will be seen that biscuit production equipment must meet a wide range of process requirements to produce the structure, texture, volume, appearance, colour and moisture content required by each product. The abilities to mix and form the doughs and to control heat transfer, baking temperature profile, time and humidity are critical factors in producing good quality products.

BIBLIOGRAPHY

Almond, N., 1989. Biscuits, Cookies and Crackers. Vol. 2. Elsevier Science Publishers Ltd., Barking, Essex, England.

Davidson, I., 2016. Biscuit Baking Technology, 2nd ed. Academic Press, London, England.

Manley, D., 1996. Technology of Biscuits, Crackers and Cookies, 2nd ed. Woodhead Publishing Ltd., Cambridge, England.

Sykes, G., 2016. Documents detailing biscuit process and recipes provided for Baker Pacific Ltd., Hong Kong, Baker Pacific Ltd., www.bakerpacific.com.hk; www.bakerpacific.net.

CHAPTER 2

Ingredient Storage and Handling

2.1 BULK DELIVERIES OF DRY AND LIQUID INGREDIENTS

For most bakeries with more than two production lines, flour and sugar may be delivered by bulk road tanker and stored in silos (Fig. 2.1).

2.1.1 Flour

Flour will be delivered directly from the tanker to the bakery silos. The delivery hose from the tanker should fit the receiving point at the factory, and the flour will be blown by an air blower on the tanker or a blower at the bakery. The tanker will normally discharge up to 30 t of flour in 20–30 min.

Most bakeries will use at least two types of flour, one 'hard' and one 'soft', which will require separate delivery and storage facilities. For some products, the two flours may be blended in the service bin before mixing (Fig. 2.2).

2.1.2 Sugar

Sugar may also be delivered by tanker, either as dry bulk sugar or liquid sugar (Fig. 2.3).

2.2 STORAGE OF DRY BULK MATERIALS

2.2.1 Flour and Sugar

Flour and sugar will be stored in silos, either external to the bakery or internal. Silos will usually be made from epoxy-coated steel and may be insulated. Stainless steel, aluminium and flexible fabric silos are also available.

External silos are common, but the bakery must be aware of the potential disadvantages (Fig. 2.4):

- The materials can be affected by significant changes in temperature and humidity.
- The rate of usage—normally, the flour or sugar should be used within a week of the delivery.

https://doi.org/10.1016/B978-0-12-815579-0.00002-7

Fig. 2.1 Flour delivery tanker from Clugston Logistics.

Fig. 2.2 Tanker discharge to flour silos from Bühler Group.

Fig. 2.3 Delivery of liquids by road tanker from Abbey Logistics Group www.abbeylogisticsgroup.com.

Fig. 2.4 Silos from Daxner Germany, www.daxner.com.

- Sugar can be more difficult to store if the particle size is small.

Silos are filled pneumatically from the top with facility for the following:

- Separation of material and conveying air
- Filtration
- Explosion relief
- Access

Silos may be discharged to the pneumatic conveying system with rotary feeders, fluid discharge units for flour or vibration units for sugar.

2.2.2 Silo Design Considerations

- Number of ingredients and usage rate (stock may be held for 1–2 weeks).
- It is usual to have more than one silo for each flour type.
- Silo can be 1.25 times the volume of the tanker delivery.
- Square or round type.

2.2.3 Operational Issues

- Cleaning of the silos
- Bridging and compacting of the materials (often due to humidity)
- Measuring the quantity of material in the silo
- Contamination of the material and eliminating metal, foreign material and infestation (Figs 2.5 and 2.6)

Fig. 2.5 Pneumatic discharger to fluidise the flour from Bühler Group.

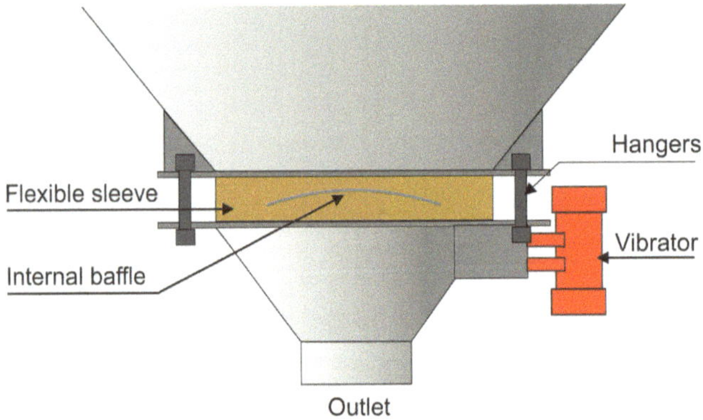

Fig. 2.6 Bin activator for difficult-to-handle materials such as sugar.

2.3 PNEUMATIC CONVEYING

The flour and sugar will be pneumatically conveyed from the silos to service bins, usually of 3–5 t capacity. The conveying system will transfer the materials from the service bins to weigh hoppers over the dough mixers as required (Figs 2.7 and 2.8).

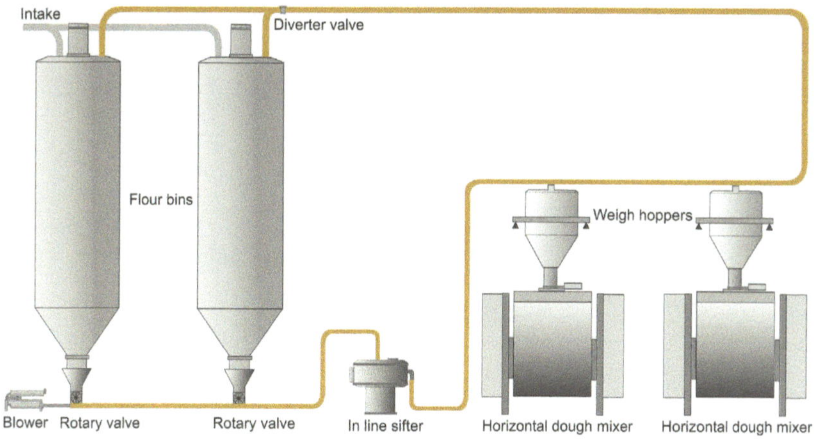

Fig. 2.7 Pneumatic conveying system to feed flour to the dough mixers.

Fig. 2.8 Pneumatic conveying of flour to the mixers.

2.4 DELIVERY OF MATERIALS IN BAGS

Flour, sugar and other dry ingredients may be delivered in bags. The material may be manually dumped from the bags to a sifter and then elevated to a weigh hopper or storage bin. The weight of material delivered in bags will be checked on a platform scale (Figs 2.9 and 2.10).

Fig. 2.9 Platform scales from Brecknell, www.brecknellscales.com.

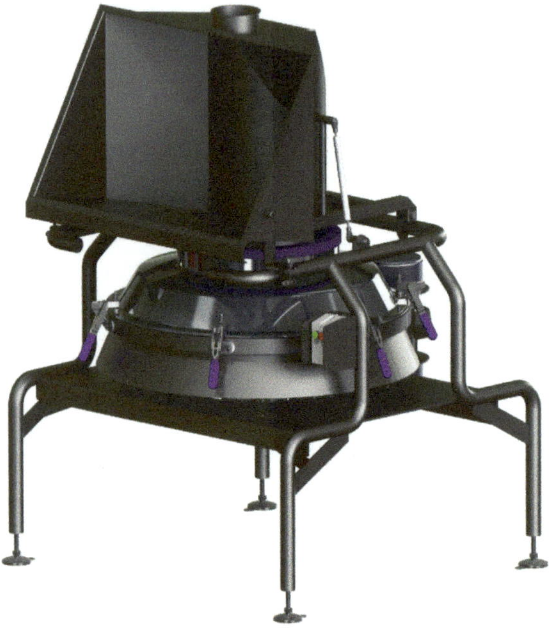

Fig. 2.10 SM EASILIFT check screener for powder and ingredients from containers or sacks, farleygreene.com.

The EASILIFT range is robust and powerful with a full dusthood protecting operators and a unique power lift system, by Farleygreene, the United Kingdom.

2.5 SIFTING OF RAW MATERIALS

Flours and other raw materials are always sifted to remove foreign matter from the mills or from bags.Vibratory sifters are available for large automatic systems or for manual feeding (Fig. 2.11).

Fig. 2.11 KTS-VS2 vibrating control screener from GKM Siebtechnik, gkm-net.de.

This screening machine is suitable for all kinds of control and safety screening with high capacities for dry and liquid products. The machine is driven by two laterally attached vibration motors. The screen body can vibrate freely by means of dislocation via precisely adjusted springs. A vertical oscillation is generated at the centre that is ideal for control screening. The fine product drops through the centre of the machine almost in free fall (Fig. 2.12).

Centrifugal flour sifter designed to provide users with an efficient dust-tight high-throughput nonvibratory sieving machine, the 'Sievmaster Rota' series, gives accurate material screening with continuous fine and coarse discharge.

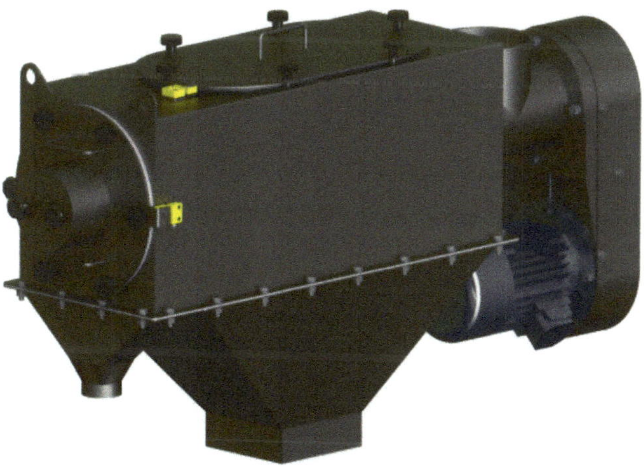

Fig. 2.12 SM ROTA series centrifugal flour sifter for high-volume screening from Farleygreene, the United Kingdom, farleygreene.com.

2.6 SUGAR MILLING

Sugar crystal size and size variation are important factors for the biscuit texture. Sugar will normally be delivered as granulated sugar with a crystal size of 450–600 μ. For some products, the sugar crystals will be milled to a fine powder of less than 150 μ (Figs 2.13–2.15).

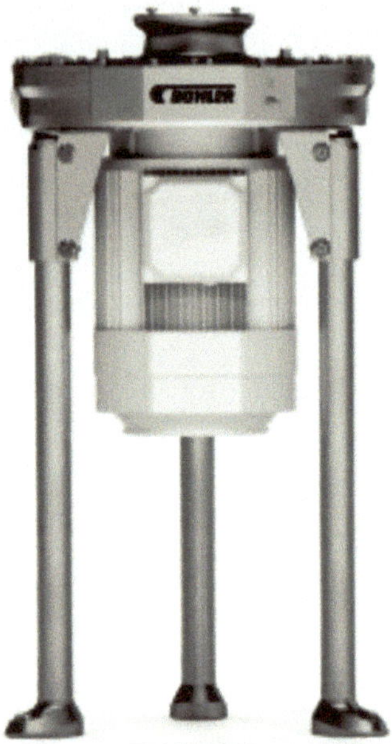

Fig. 2.13 Industrial sugar mill: Bühler MJB-22 with graduated crystal size down to 140 μm and output of 0.5–3.0 t/h.

Fig. 2.14 Fine grinder for sugar from Prater Industries, www.praterindustries.com.

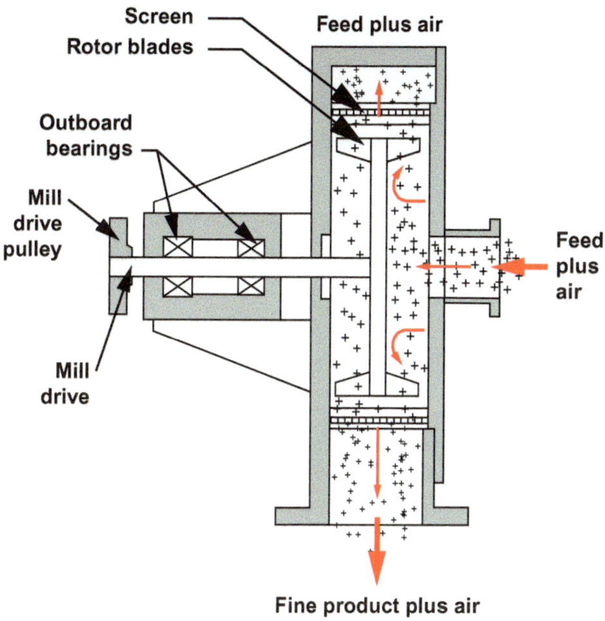

Fig. 2.15 Details of fine grinder for sugar from Prater Industries.

2.7 MATERIAL RECOVERED FROM PRODUCTION

Many biscuits contain rework such as damaged biscuits recovered from production. Hygiene in collection, storage and processing is essential. The biscuits are ground and may then be added to some recipes at up to 10% of the total.

2.8 SHORTENING

Shortening may be purchased as hydrogenated plasticised fats, which are refrigerated and may be fed directly to the mixer, or melted and fed as an oil. The typical temperature for use is 27°C at which the fat has an SFI of 14% (Figs 2.16 and 2.17).

2.9 STORAGE OF LIQUID MATERIALS

Liquid materials are stored in tanks, normally of stainless steel construction. These may be heated and insulated to maintain the material at a temperature where it is liquid and free flowing. The liquids are pumped to the mixer with a metering system (Figs 2.18–2.21).

Fig. 2.16 Vegetable shortening from Grüninger AG, Switzerland.

Fig. 2.17 Vegetable shortening from Erapoly, Malaysia.

2.10 DELIVERY OF INGREDIENTS TO THE MIXERS

The main ingredients will be delivered to the mixers via weigh hoppers and metering systems for water and other liquid ingredients. Weigh hoppers above the mixers are mounted on load cells, and the system operates by measuring the 'weight loss'. A slide valve below the hopper will open to allow the material to be dropped into the mixer bowl. When the correct weight of material has been delivered, the automatic slide valve will close the port below the hopper (Figs 2.22 and 2.23).

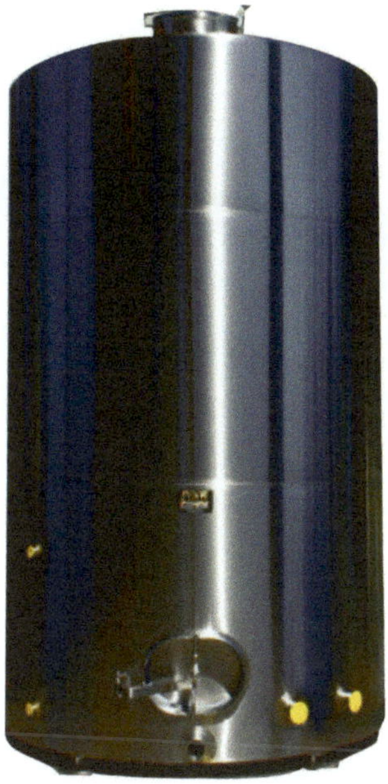

Fig. 2.18 Edible oil tank from Jiangsu Prettech Machinery and Technology.

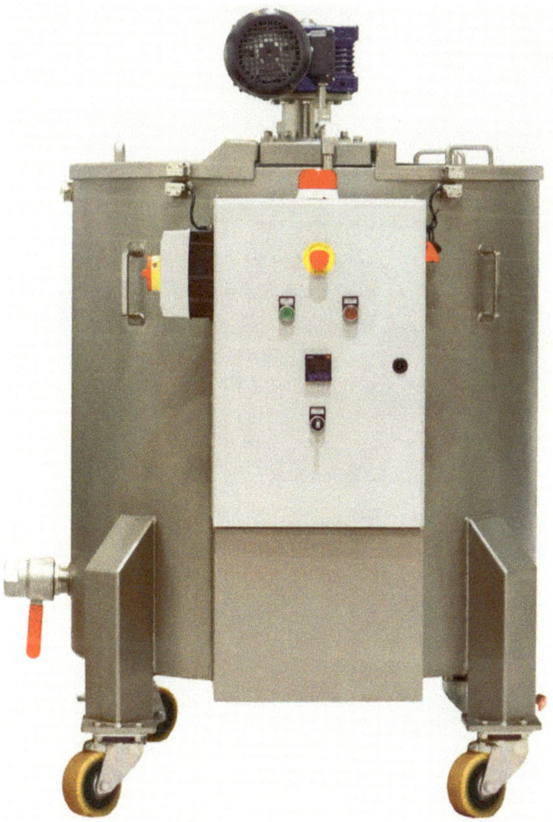

Fig. 2.19 Chocolate storage tank with heated water jacket, adjustable temperature control and stirrer from Loynds International Ltd. Capacities 75–5000 kg, www.loynds.com.

Fig. 2.20 Albany Pumps GJ gear pump: fast maintenance and hygienic stainless steel pump for the food industry, albany-pumps.co.uk.

Fig. 2.21 Bell Flow Systems batch water meter.

Fig. 2.22 Mixer feed systems from Dingson Food Machinery.

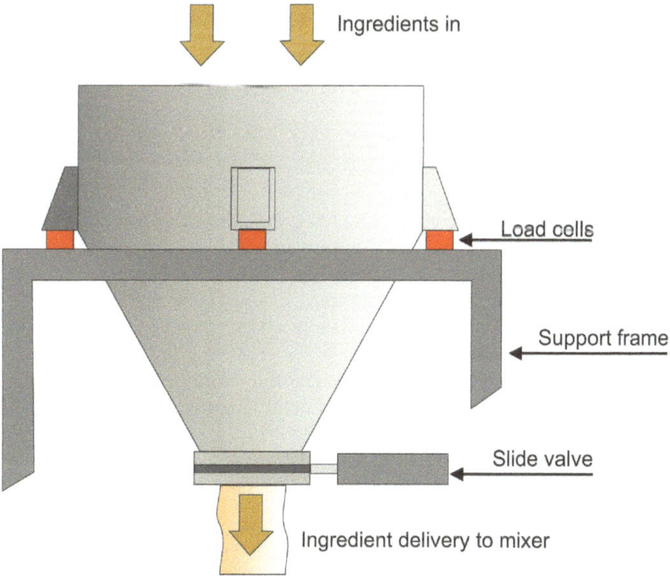

Fig. 2.23 Weigh hopper for dry ingredients.

Fig. 2.24 MBS scales from Brecknell.

2.11 SMALL INGREDIENTS

2.11.1 Colours, Flavours, Additives

A variety of materials in both liquid and powder form that will be stored at constant temperatures as required. These will generally be weighed and placed in small containers for feeding to the mixer (Fig. 2.24).

2.11.2 Boxed Ingredients

Various small ingredients such as raisins and fruit will be packed in cartons and will require inspection to remove stones, stalks and other foreign material.

BIBLIOGRAPHY

Almond, N., 1989. Biscuits, Cookies and Crackers. vol. 2. Elsevier Science Publishers Ltd.
Albany Pumps, 2017. https://www.albany-pumps.co.uk/.
AZO GmbH + Co. KG, 2016. www.azo.com.
Baker Perkins Ltd, 2016. www.bakerperkins.com.
Bell Flow Systems Ltd, 2016. www.bellflowsystems.co.uk.
Bühler Group, 2016. www.buhlergroup.com.
Brecknell: Avery Weigh-Tronix, 2017. www.averyweigh-tronix.com.
Clugston Logistics, 2017. https://www.clugston.co.uk/logistics/.
County Scales Ltd, 2017. www.countyscales.co.uk.
Daxner International Bulk Solids Technology, 2017. www.daxner-international.com.
Dingson Food Machinery Ltd, 2016. www.dsm-mc.com.
Erapoly Global Sdn Bhd, 2016. www.erapoly.com.
Farleygreene, 2017. www.farleygreene.com.
Gericke Ltd, 2016. www.gericke.net.
GKM Siebtechnik, 2017. https://gkm-net.de.
Grüninger AG, 2016. www.grueninger.ch.

Guerin Systems, 2016. www.guerinsystems.com.
Jiangsu Prettech Machinery & Technology Ltd, 2017. www.sinoprettech.com. www.prettechtank.com.
Kemutec, 2017. www.kemutecusa.com.
Lianhe Tech, 2016. www.lianhetech.com.
Manley, D., 1996. Technology of Biscuits, Crackers and Cookies, second ed. Woodhead Publishing Ltd.
Manley, D., 1998. Ingredients. Woodhead Publishing Ltd.
Mazzetti Renato S.p.A, 2016. www.mazzettirenato.it.
Prater Industries, 2017. https://www.praterindustries.com.
Russell Finex, 2016. www.russellfinex.com.
Selis, 2016. www.selis.com.tr.

CHAPTER 3

Dough Mixing

3.1 THE MIXING PROCESS

There are several processes that occur during the mixing of the dough. These include the following:

- Dispersion of all the ingredients and blending into a homogeneous mass
- Hydration of the flour
- Emulsification of the fats and water
- Development of the gluten formed by the proteins in the flour
- Activation of leavening agents

3.1.1 Crackers and Semi-Sweet Doughs (Fig. 3.1)

Fig. 3.1 Crackers and semi-sweet biscuits.

Crackers and hard doughs have relatively high amounts of water and small amounts of fat and sugar. During the mixing process, the gluten is developed by vigorous cutting and shearing of the dough. This results in an extensible dough suitable for sheeting and cutting to form the dough pieces. The mixing action imparts work to the dough and results in an increase in temperature. In order to ensure consistency of dough batches, the dough for products such as semi-sweet biscuits is mixed to a fixed preset temperature.

For most products, the process is an 'all-in' mix, when all the ingredients are fed to the mixer at the start. After mixing, cracker doughs will be fermented.

Biscuit, Cookie and Cracker Production
https://doi.org/10.1016/B978-0-12-815579-0.00003-9

For some crackers, for example, soda crackers, the mixing process is in two stages, a 'sponge' and then a second mix with additional ingredients. The initial mixing is with approximately 2/3 of the flour quantity and half of the fat. The initial mix results in a 'sponge' that is fermented for 12 h. The remainder of the flour and fat is then added, with soda and salt, and the dough is then mixed and then fermented for a further 12 h. The process is known as 'sponge and dough'.

3.1.2 Short Doughs (Fig. 3.2)

Fig. 3.2 Short-dough biscuits.

In mixing short doughs, it is important that the gluten is not developed. The mixing is therefore in two stages. The first stage is called a 'cream up'. During this stage, the fat, sugar, water and small ingredients are mixed to obtain a soft mass. The flour is then added, and the final stage of the mix is as short as possible so that the gluten web is not developed and the resulting dough will have a short texture. After mixing, the dough will be stood for approximately 30 min to allow the hydration of the flour before forming.

3.1.3 Cookie Doughs (Fig. 3.3)

Fig. 3.3 Cookies.

Cookie doughs have high fat and sugar. The doughs are soft, and they are formed by extrusion and depositing directly onto the oven band. As for short doughs, the mixing process is in two stages, with a 'cream' stage, followed by adding the flour and mixing for a short time to prevent gluten development. The doughs are kept cool and usually mixed to a temperature of around 20°C.

3.1.4 Deposited Cake Batters, Creams (Fig. 3.4)

Fig. 3.4 Biscuit creams and deposited Jaffa Cake.

Doughs for snack cakes, soft cookies and biscuit creams may be mixed on a planetary mixer that provides a whisk and aeration to give a very light, low-density fluid dough for depositing and creams suitable for sandwich biscuits.

3.2 TYPES OF MIXER

There are five main types of mixer, each with advantages for particular products and applications: vertical spindle mixers, high-speed horizontal mixers, horizontal mixers for cookie doughs, continuous mixers and planetary mixers.

3.2.1 Vertical Spindle Mixers

These mixers use a mobile dough tub into which the ingredients are fed manually or automatically from hoppers. The dough tub is then taken to the mixer and either the vertical spindles are lowered into the tub or the tub is lifted to the spindles.

Vertical spindle mixers are widely used for fermented crackers, particularly soda crackers. The sponge and dough are mixed in the dough tub, and after each mix, the tub is then transferred to the fermentation room. After fermentation of the sponge, the dough tub is brought back to the mixer for the final stage of the mixing cycle. In a two-stage mixing process for soda crackers, additional ingredients are added before the final mix.

The mixing action is slow and thorough and generates little heat. Hard doughs may require a total mixing time of 60–90 min to develop the gluten sufficiently (Figs 3.5–3.7).

Fig. 3.5 T.L. Green 3 spindle mixer with independently driven spindles from Reading Bakery Systems. Capacity up to 1000kg.

Fig. 3.6 Vertical mixer with three spindles and 1000kg capacity from Dingson Food Machinery.

Fig. 3.7 Vertical spindle mixer from Apinox Srl. These pictures belong to Apinox Srl—Italy. This mixer has been specifically designed for the production of soda cracker and cream cracker with two stages, fermented dough and delicate dough.

3.2.2 Horizontal High Speed Mixers

This type of mixer is very widely used for a variety of snack crackers, semi-sweet doughs and short doughs.

The main dry ingredients may be fed from an automatic weigh feeding system mounted above the mixer. Water and other liquid ingredients may be metered and fed automatically. Typically, small ingredients are fed to the mixer bowl in a half-tilt position by hand but can also be fed automatically from a small hopper above the mixer bowl.

The mixer bowls are stainless steel with a water jacket through which cold water is circulated. For some snack products, a heated water jacket may be used to gelatinise potato starches.

The bowl tilts to discharge the mixed dough. The tilt may fully invert the mixer bowl to allow automatic discharge of dough directly to a hopper below the mixing floor for feeding the sheeter of the forming equipment.

Baker Perkins mixer capacities range from 450 to 1100 kg. Two speed machines have timers to set times at slow speed (30 rpm) and then high speed (60 rpm). Alternative drives provide variable speed control.

The mixer blade provides a vigorous extruding and shearing action to develop the gluten web, and this action results in an increase in temperature. The dough temperature is monitored by a thermocouple in the wall of the mixer bowl (Figs 3.8 and 3.9).

Fig. 3.8 Baker Perkins high-speed horizontal mixer.

Fig. 3.9 Baker Perkins shaftless mixer blade.

The shaftless mixer blade design gives an end-to-end mixing action to incorporate all ingredients evenly including those fed in small quantities. It avoids dough adhering to a central shaft, which will result in it not being fully incorporated into the dough mass.

Horizontal mixers from Spooner Vicars have a helical 'Sprag' mixer blade, which allows the temperature sensor to be incorporated in the centre of the dough mass.

3.2.3 Horizontal Mixers for Cookie Doughs

Sigma arm mixers have been an industry standard for mixing cookie and biscuit doughs for many years. Mixers have stainless steel construction, fully tilting bowl with water jacket for temperature control and two speed drives or variable frequency drive (Fig. 3.10).

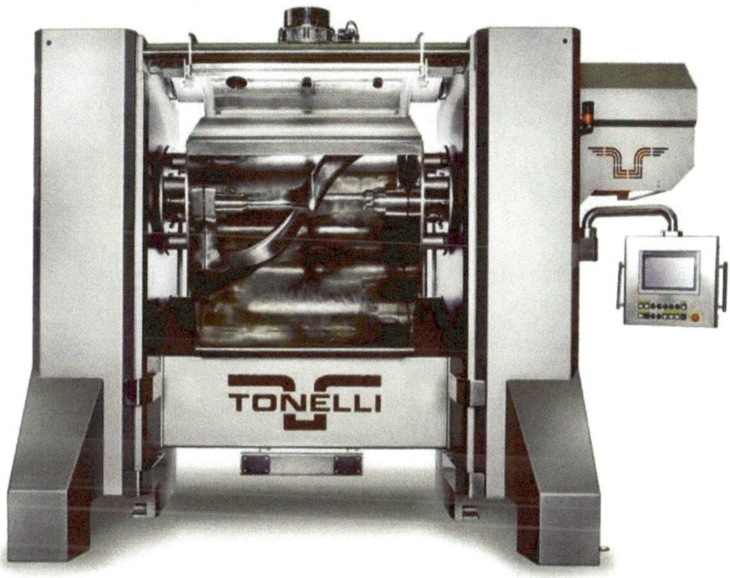

Fig. 3.10 Tonelli TR series mixer.

3.2.4 Continuous Mixers

Continuous mixers have been successfully used for a wide range of cracker, biscuit, and cookie doughs. They are supplied with fully automatic ingredient feed systems and usually for single-purpose production lines (Fig. 3.11).

The LDX continuous mixer combines all minor ingredients and a portion of the required flour in the first mixing stage. In this stage, special mixing elements cut the fat into the other ingredients. In the final stage, the remaining flour is added to create the dough. This mixer is ideal for most types of cookies or any application where ingredients are blended before flour is added.

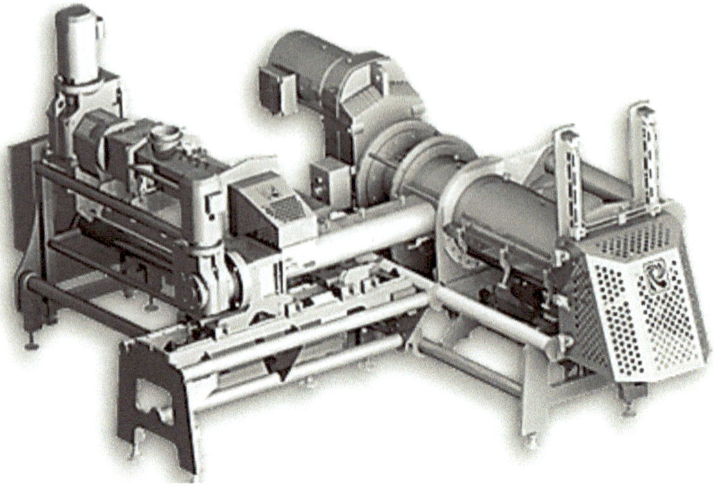

Fig. 3.11 Reading Bakery Systems LDX continuous mixer.

3.2.5 Planetary Mixers (Fig. 3.12)

Fig. 3.12 'Planetaria' mixer from Tonelli SpA.

The mixing bowls for planetary mixers are mobile, and ingredients are fed remotely.

Planetary mixers have interchangeable tools suitable for a variety of applications, including cookies, cakes and creams. The planetary mixing system uses one or two mixing tools with a scraper. These are rotated on a preset planetary system to suit the product. The mixing speed of the tools and the rotation are separately driven.

Planetary mixers may have a facility to inject air into the mix to create a light, low-density cake batter.

The dough may be transferred by pump or by lifting and tilting unit to feed the hopper of a depositing machine from the mixing bowl.

BIBLIOGRAPHY

Apinox S.r.l, 2016. www.apinox.it.
Baker Perkins Ltd, 2016. www.bakerperkins.com.
Dong Yang Food Machinery, 2016. www.mixer.co.kr.
Middleby Corporation, 2016. www.middleby.com.
Oakes, E.T., 2016. www.oakes.com.
Peerless Food Equipment, 2017. www.peerlessfood.com.
Reading Bakery Systems, 2016. www.readingbakery.com.
Spooner Vicars, 2017. www.spoonervicarsbakery.com. www.middprocessing.com.
Tonelli Group Spa, 2017. www.tonelli.it.
Green, T.L., 2016. www.readingbakery.com.

CHAPTER 4

Dough Feed Systems

4.1 DOUGH FEED SYSTEMS

After mixing the dough, it is fed from the dough tub to the forming machines. Most horizontal high-speed mixers now have tipping angle of over 100°, which allows them to automatically discharge the dough into a dough feed system on a floor below the mixing room. Mixers with removable bowls require the bowls to be manually moved to a lift and tilt unit to discharge the dough into a sheeting machine.

4.1.1 Dough Tub Lift and Tilt Units

For doughs that will be sheeted to feed a biscuit cutting machine line or a laminator, a tub lift and tilt unit may be used to tip the bulk dough into the hopper of the sheeter (Figs 4.1 and 4.2).

Fig. 4.1 Lift and tilt units to feed the sheeter hopper from a dough tub.

Biscuit, Cookie and Cracker Production
https://doi.org/10.1016/B978-0-12-815579-0.00004-0

Fig. 4.2 Bulk dough being tipped from a lift and tilt unit to the hopper of a sheeter.

4.1.2 Dough Feed Conveyor With Metal Detector

Alternatively, for doughs for crackers and semi-sweet biscuits, the bulk dough from the mixer can be metered by a two-roll feeder and then conveyed to the sheeter of the biscuit cutting line or laminator. The conveyor should have a metal detector and reject system to avoid any metal being delivered to the sheeter rolls.

4.1.3 Live Bottom Bin Dough Feeder

An alternative feed system uses a 'live bottom bin'. The dough is tipped from the tub into the hopper of the live bottom bin. The base of the hopper is a conveyor that will move the dough mass forward, where it is cut into short pieces by a vertical guillotine and delivered onto the dough feed conveyor. The dough feed conveyor delivers the dough to the sheeter of the biscuit cutting line or laminator. The conveyor is controlled by a level sensor in the hopper of the forming machine and operates intermittently to meter the dough as required by the forming line.

Some systems use tines or a 'chunker' with rolls in place of the guillotine. These systems divide the dough into separate pieces for conveying to the forming line (Figs 4.3–4.5).

Fig. 4.3 Dough tub tilt unit to feed a live bottom bin.

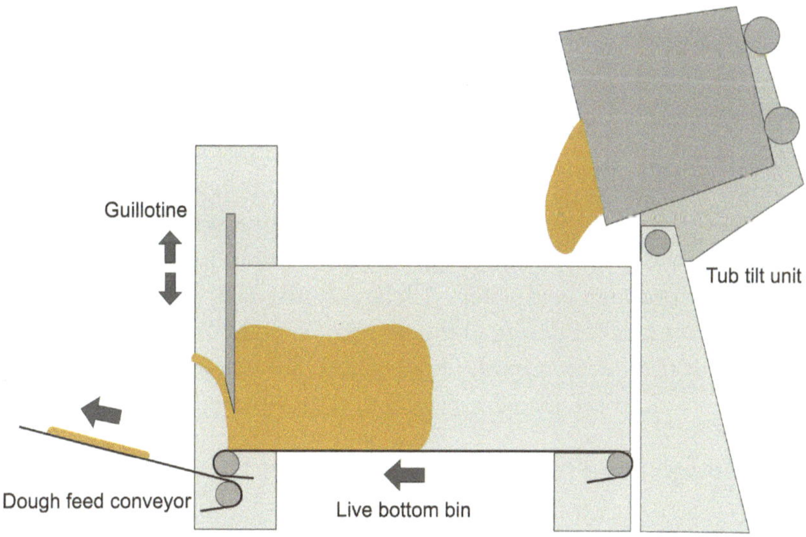

Fig. 4.4 Live bottom bin arrangement.

Fig. 4.5 Automatic dough feed system from Arrow Design Services with options for adjustable dough gate from live bottom feeder, sheeting rolls, kibbler and chunker arrangements.

4.1.4 Dough Feed to a Rotary Moulder

Rotary moulders must be fed very evenly and at a constant rate to ensure accuracy in the weights of the moulded dough pieces. The bulk dough may be fed from a two-roll sheeter or a live bottom bin to the dough feed conveyor. The dough is then kibbled, which is broken into small pieces, above the hopper of the rotary moulder and delivered very evenly across the width of the machine. The hopper of the rotary moulder will have level probes, and the dough feed system will be automatically controlled to maintain a constant head of dough in the hopper. Vertical guides may be used in the chute to the rotary moulder hopper to ensure an even feed of dough across the width of the machine (Fig. 4.6).

4.1.5 Dough Feed to Depositing Machines

Depositing doughs (or batters) are relatively fluid and may be pumped into the depositor hopper. Alternatively, they may be poured from a hopper above the depositor with an adjustable gate to control the flow of batter (Fig. 4.7).

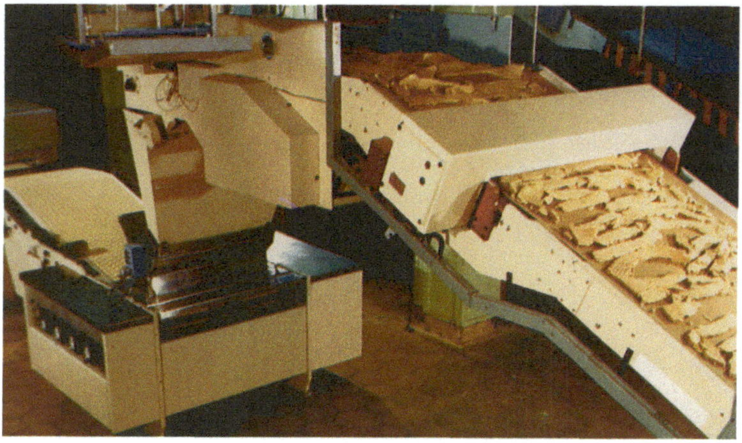

Fig. 4.6 Dough feed to a rotary moulder with metal detector.

Fig. 4.7 Cake depositor fed with batter from an aerating mixer (on the right) by pump.

BIBLIOGRAPHY

Arrow Design Services Ltd, 2017. www.arrowdesignservices.co.uk.
Baker Perkins Ltd, 2016. www.bakerperkins.com.
Dingson Food Machinery, 2016. www.dsm-mc.com.
Egan Food Technologies, 2017. www.eganfoodtech.com.
Food Processing Equipment s.r.l, 2017. www.fpe2000.it.
Spooner Vicars, 2017. www.spoonervicarsbakery.com. www.middprocessing.com.

CHAPTER 5

Dough Piece Forming: Biscuit Cutting Machine

5.1 CUTTING MACHINE UNITS

Forming a continuous dough sheet and then cutting the dough pieces to the required size and shape and printing the design on the surface of the dough pieces is a traditional and common method of forming the dough for a wide variety of biscuits. These include all semi-sweet biscuits, crackers and hard biscuits (Fig. 5.1).

Fig. 5.1 Biscuits formed on a cutting machine.

The process involves feeding the bulk dough to a sheeting line that has typical working widths of 1000, 1200 and 1500 mm. The cutting machine line comprises the following machines (Fig. 5.2):

1. Three (or four)-roll sheeter to which the bulk dough is fed. The sheeter produces the initial dough sheet.
2. A series of gauge roll units comprising pairs of heavy rollers to reduce the thickness of the dough sheet. Each gauge roll unit will reduce the thickness of the dough sheet by approximately 50%. The final gauge roll unit gauges the dough sheet to the correct thickness for cutting (typically 0.75–2.5 mm).

Biscuit, Cookie and Cracker Production
https://doi.org/10.1016/B978-0-12-815579-0.00005-2

3. Relaxation conveyor to reduce the stress in the dough sheet caused by the action of the gauge rollers. This stress will cause the dough pieces to shrink excessively during baking, if it is not relieved before cutting.
4. Biscuit cutting conveyor that supports and carries the dough sheet below the cutting rolls.
5. Biscuit cutting machine with a single cutting roll or separate printing and cutting rolls. An alternative system, not often used now, is a vertical stamping machine for printing and cutting the dough pieces. This system may be employed for doughs with inclusions such as currants.
6. Scrap lift and return conveyors. These conveyors lift the scrap dough as a lattice from around the cut dough pieces and return the scrap to the sheeter that will incorporate it in the dough sheet.
7. Decorating conveyor for the application of egg wash or other materials.
8. Panning conveyor that deposits the dough pieces onto the oven band.
9. Sugar and salt sprinklers with recovery system.

Fig. 5.2 Dingson Food Machinery cutting machine line with three-roll sheeter, three gauge roll units, rotary cutter and panner. Side scrap return conveyor to sheeter.

5.2 DOUGH SHEETING MACHINES

For most applications, a three-roll sheeter is used. This is designed to produce a continuous, consistent dough sheet with the correct width. The bulk dough is fed directly to the hopper of the sheeter. A grooved forcing roll feeds the dough between a pair of gauging rolls that produce the sheet of dough and lay it onto a conveyor that will carry the dough sheet to the first gauge roll unit.

Scrap dough from the cutter is also fed to the hopper of the sheeter, usually at the back of the hopper so that when sheeted the scrap dough is on the base of the dough sheet (Fig. 5.3).

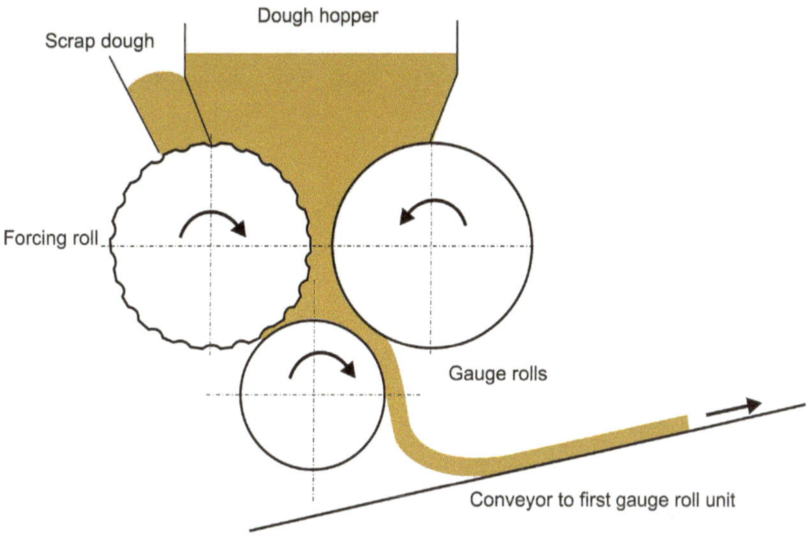

Fig. 5.3 Diagram of three-roll sheeter.

The gaps between the forcing roll and the top gauge roll and between the two gauge rolls are adjustable. Usually, the forcing roll gap is approximately twice the gauge roll gap (Fig. 5.4).

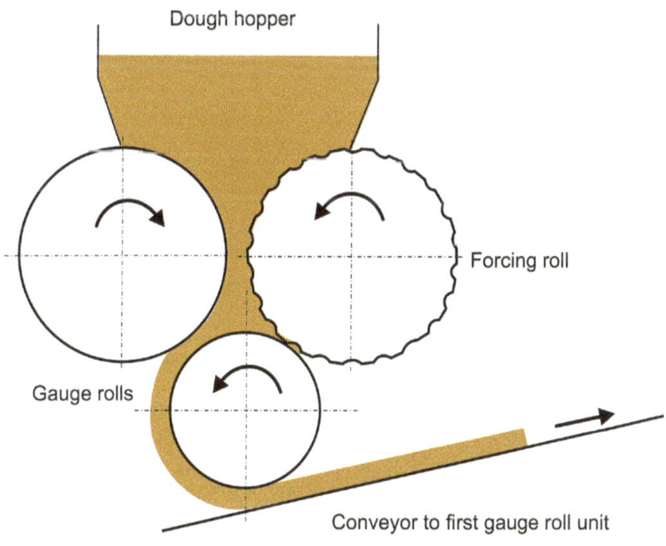

Fig. 5.4 Back discharge sheeter.

An alternative configuration of the sheeter, known as a back discharge sheeter, is often used to feed softer doughs or to feed a vertical laminator.

5.3 GAUGE ROLL UNITS

Sheeters will produce a dough sheet for semi-sweet biscuits with a thickness usually around 9.0mm. This dough sheet now needs to be reduced to the correct thickness of around 1.3mm for cutting the dough pieces. This reduction is achieved by a series of gauge rolls that reduce the dough thickness step by step. Each reduction is approximately 50%. Typical roll gap settings will be as follows:

Forcing roll gap on sheeter	18.0mm
Gauging gap on sheeter	9.0mm
First gauge roll	5.7mm
Second gauge roll	2.5mm
Final gauge roll	1.1mm (cutting thickness: 1.3mm)

Gauge rolls diameters are 300–420mm and are usually constructed from cast iron. They may be of solid construction and may have a bonding of stainless steel on the surface. The object is to ensure the minimum deflection under load to give a very accurate thickness of dough across the width of the dough sheet. Any variation will be evident in the weight and colour of the final baked products.

The gauge roll gaps are adjustable, and adjustable scrapers are fitted to top and bottom gauge rolls. The bottom roll will have a slightly faster speed, so that the dough will tend to follow this roll and not adhere to the top roll (Figs 5.5 and 5.6).

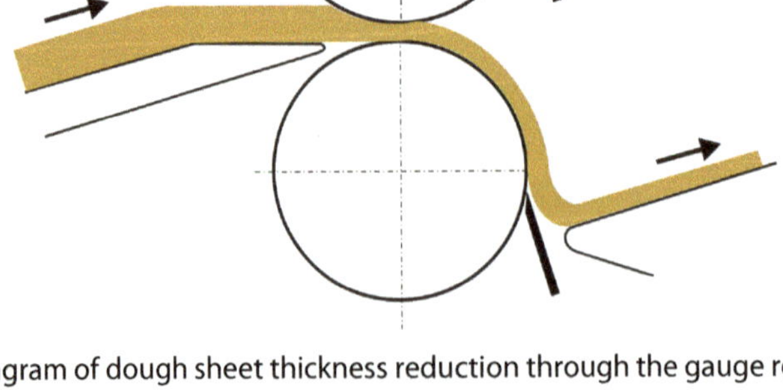

Fig. 5.5 Diagram of dough sheet thickness reduction through the gauge roll unit.

Fig. 5.6 Gauge roll units.

5.4 RELAXATION CONVEYOR

The process of reducing the dough thickness by a series of gauge rolls stresses the dough structure, and this stress must be relieved; otherwise, the dough pieces will shrink during baking, and the result will be misshapen biscuits. The relaxation conveyor, usually 2.0 m long, runs more slowly than the delivery of the dough sheet from the final gauge roll unit. The dough sheet will ripple as it feeds on to the relaxation conveyor, and as the stress is relieved, the ripples will diminish (Fig. 5.7).

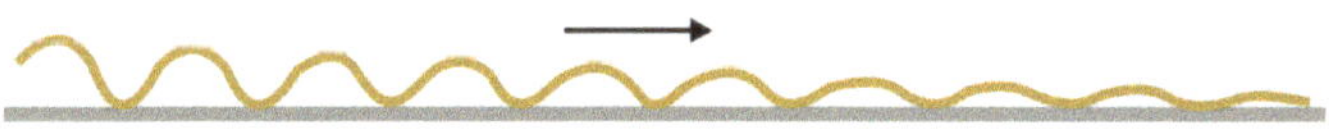

Fig. 5.7 Dough sheet ripples on a relaxation conveyor.

5.5 DOUGH PIECE CUTTING AND SCRAP DOUGH RETURN

The dough sheet is supported on the cutting web as it passes under the cutting rolls. The web may be of woven cotton or a synthetic web. It is important that the dough adheres satisfactorily to the web during the cutting process and does not pick up on the rolls.

A rotary cutting machine employs one or two rolls. The two-roll machine is preferred for most products as it allows the printing and dockering to be adjusted separately from the cutting of the dough pieces. The pressure of each roll must be accurately adjusted to ensure a clear print of the biscuit design and a clean cut of the dough pieces. Under the cutting web, directly below the cutting and printing rolls are support rollers. These are adjustable to ensure the correct cutting and printing pressure. The rotary cutter also has a phasing adjustment to ensure that the print is exactly in the centre of the cut dough pieces (Figs 5.8–5.10).

The scrap dough lattice is carried up to cutting discs that cut the lattice into small sections. These are dropped onto a cross conveyor that will deliver the scrap to a return conveyor at the side of the cutting machine. The scrap dough will be returned to the sheeter, where the scrap will be incorporated with the new dough to produce the dough sheet.

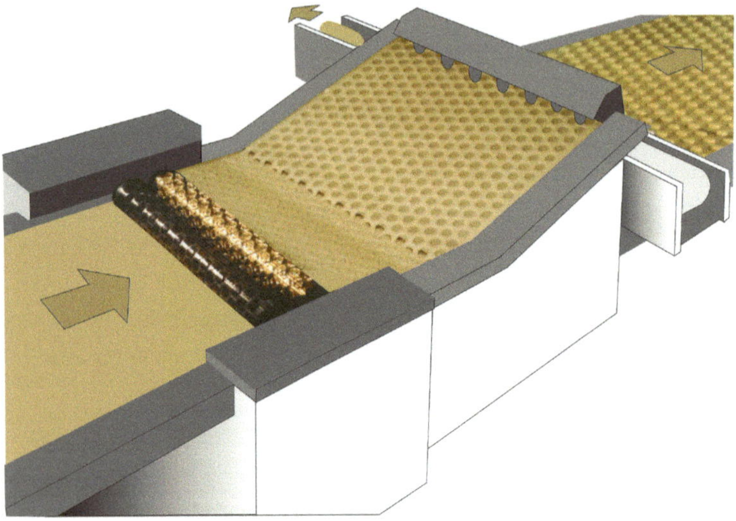

Fig. 5.8 Two-roll rotary cutter with print and docker roll and bronze cutting roll. The scrap dough lattice is lifted and cut and dropped onto the cross scrap conveyor. The cut dough pieces are conveyed towards the panner and the oven.

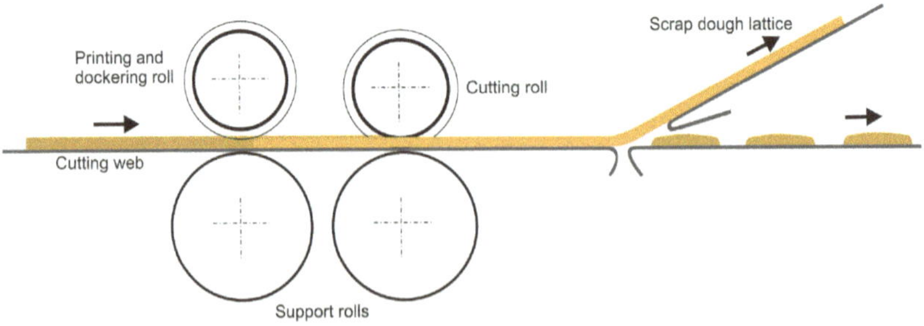

Fig. 5.9 Rotary cutter with separate printing and cutting rolls and scrap dough pick up.

Fig. 5.10 Baker Perkins rotary cutter with scrap lift and cross conveyor.

5.5.1 Biscuit Cutting and Moulding Rolls

Printing and cutting rolls are engraved from a bronze roll or more frequently now have plastic inserts mounted on a steel roll. Some snack and cracker products are cut without scrap and baked in a sheet (Figs 5.11 and 5.12).

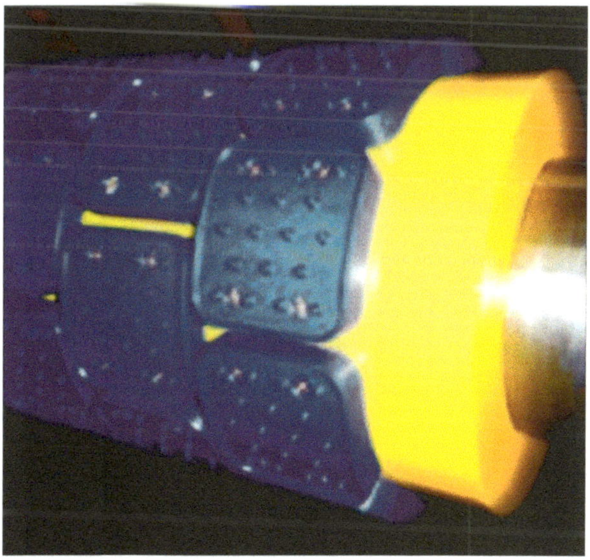

Fig. 5.11 Cutter roll with plastic inserts.

Fig. 5.12 Scrapless cutting roll.

Cutting rolls from Errebi Technology <errebi.net>.

The design of the plastic inserts is critical to achieve the best final product. The insert design below for Maria is by Errebi Technology (Figs 5.13–5.15).

Fig. 5.13 Maria biscuit.

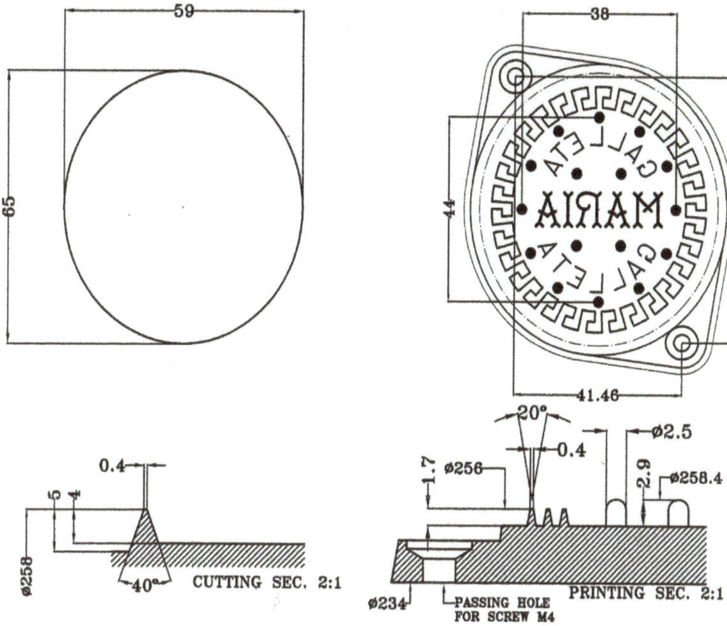

Fig. 5.14 Design for inserts for a cutting roll by Errebi Technology.

Reciprocating cutters may still be used for products with inclusions such as currants. The cutting head of the machine drops vertically onto the dough sheet to print and cut as it follows the motion of the conveyor. It then rises and returns to print and cut the following rows.

5.5.2 Scrap Dough Return to the Sheeter

Fig. 5.15 Scrap dough pick up and transfer to a cross conveyor and then to a side scrap return conveyor. The scrap dough is conveyed back to the dough sheeter. Biscuit cutting machine from Baker Perkins.

5.6 DECORATING CONVEYOR

The cutting web may deliver the dough pieces onto a separate decorating conveyor, where the surfaces of the dough pieces may be gilded with a liquid egg or milk solution to provide a rich golden gloss to the biscuit. Other materials such as sugar, salt or seeds may be sprinkled on to the dough pieces (Fig. 5.16).

Fig. 5.16 The decorating conveyor follows the cutting web and the scrap lift and cross conveyor. This line has a 'washover' unit for applying an egg or milk solution to the dough pieces.

The decorating conveyor delivers the dough pieces to the panner, which lays the dough pieces onto the oven band.

Salt and sugar may be sprinkled onto the dough pieces before baking. This may be done on a decorating conveyor or on the panning conveyor with wire-mesh conveyor band. The wire-mesh band allows surplus salt and sugar to drop through the mesh to a recovery hopper (Fig. 5.17).

Fig. 5.17 Salt/sugar sprinkler mounted over a wire-mesh recovery conveyor from Dingson Food Machinery.

5.7 PANNER

The dough pieces are transferred to the oven band by a panner (panning conveyor) (Fig. 5.18).

Fig. 5.18 Panning conveyor delivering the dough pieces onto the oven band.

It is important that the dough piece pattern fills the oven band evenly and close to the edges of the band. If too much of the oven band is exposed at the edges, it will overheat and cause the colour of the edge biscuits to be darker.

In order to ensure that the dough pieces can be laid close to the band edges, the panning conveyor is mounted on a swivel pivot. Guide wheels are attached at each side at the delivery end nosepiece of the panner, and these engage and follow the oven band. As the oven band tracks from side to side, the panner will follow and will lay the dough pieces accurately across the width of the band.

The relative speeds of the panner and the oven band will determine the spacing of the rows of dough pieces, and this can be adjusted to achieve the optimum spacing in the oven (Fig. 5.19).

5.8 CONTROL SYSTEM

Cutting machines are generally controlled by a PLC system with an HMI and touchscreen.

Conveyor belts will be automatically tensioned and tracked.

The rolls and conveyors for the complete cutting machine line are driven by AC motors with inverter speed control. These drives operate on a

Fig. 5.19 The panner nosepiece is supported at each side on rollers that allow it to swivel from side to side, following the tracking of the oven band.

cascade system, usually with the cutting web speed as the master. When the cutting web speed is increased or decreased, the speeds of all the preceding units will follow the increase or decrease to maintain a smooth uninterrupted flow of the dough sheet.

In addition, individual speed controls for each unit are used to trim speeds to ensure the optimum flow of the dough sheet through the successive units of the cutting machine.

Roll gap adjustments are made by servomotors with PLC and HMI touchscreen control.

The roll gaps and speed settings for each product will be different and will be controlled by recipe settings in the PLC (Fig. 5.20).

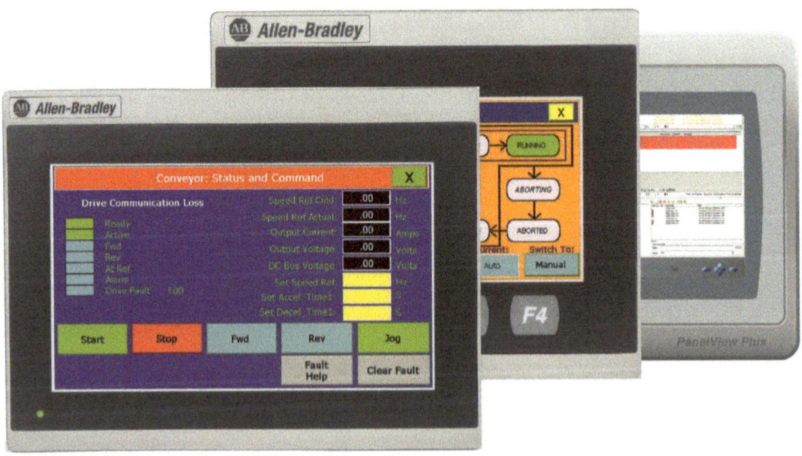

Fig. 5.20 Allen Bradley PanelView 800 and PanelView Plus screens from Rockwell Automation.

BIBLIOGRAPHY

Almond, N., 1989. Biscuits, Cookies and Crackers. vol. 2. Elsevier Applied Science.
Baker Pacific Ltd, 2017. www.bakerpacific.com.hk. www.bakerpacific.net.
Baker Perkins Ltd, 2017. www.bakerperkins.com.
Dingson Food Machinery Ltd, 2017. www.dsm-mc.com.
Laser srl., 2017. www.laserbiscuit.com.
Manley, D., 1998. Biscuit Dough Piece Forming, Manual 3. Woodhead Publishing Ltd.
Reading Bakery Systems, 2017. www.readingbakery.com.
Rockwell Automation, 2015. www.rockwellautomation.com.
Spooner Vicars Bakery Systems, 2017. www.spoonervicarsbakery.com.

CHAPTER 6

Dough Piece Forming: Laminating

6.1 DESIGN AND CONFIGURATION

The dough sheets for some crackers and hard sweet biscuits are laminated (layered) before being delivered to the gauge rolls and cutting machine. The layered structure of the dough sheet produces the open flaky structure typical of some crackers (Fig. 6.1).

Fig. 6.1 Crackers formed with laminations.

The laminating equipment consists of a three-roll sheeter, followed by gauge roll units (usually two) and a layering section. The laminated dough sheet is fed to the cutting machine line at right angles to reduce the stresses in the final dough sheet before cutting.

The laminator may be a 'horizontal laminator' where the units are installed on the floor at right angles to the cutting machine line. Because of the extensive space required, these have largely been replaced by 'vertical laminators' or 'in-line horizontal laminators'. The sheeter and gauge roll units of vertical laminators are arranged vertically above the layering section (Fig. 6.2 and 6.3).

The laminations may be made by folding the dough sheet and laying successive layers as the conveyor carries the dough sheet forward. This system gives additional stress to the edges of the dough sheet where the folds occur. To avoid this, most laminators now cut the dough sheet before laying separate sheets on top of each other (Fig. 6.3).

Biscuit, Cookie and Cracker Production
https://doi.org/10.1016/B978-0-12-815579-0.00006-4

Fig. 6.2 Baker Perkins forming a line from right to left: vertical laminator with scrap return, three gauge roll units, relaxation conveyor, rotary cutter, scrap lift and return.

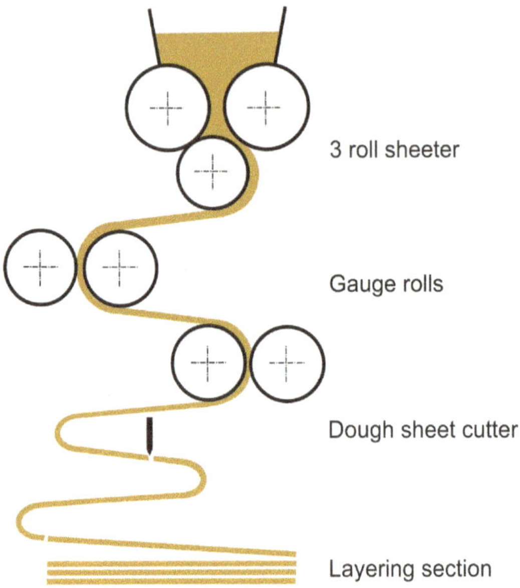

Fig. 6.3 Vertical cut sheet laminating system.

In-line horizontal laminators form the dough sheet with a three-roll sheeter and gauge roll units in line with the cutting machine. The dough sheet is then turned at right angles over a roller and then layered at right angles to the cutting machine (Figs 6.4–6.6).

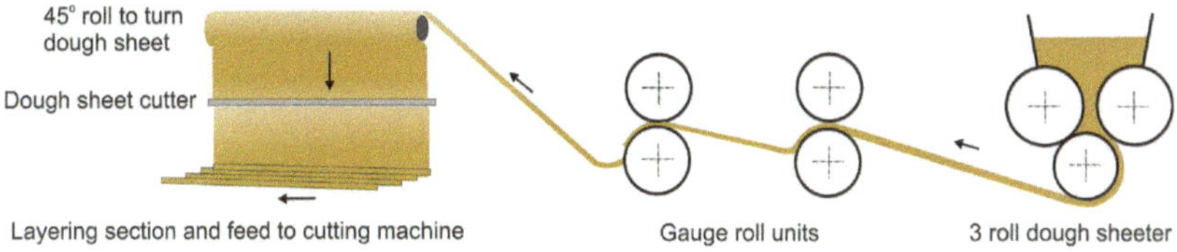

Fig. 6.4 In-line horizontal cut sheet laminator.

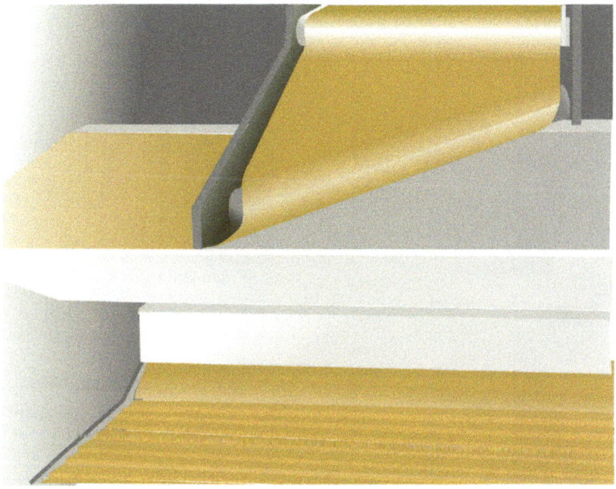

Fig. 6.5 In-line horizontal laminator cutting and layering section.

Fig. 6.6 Layered dough sheets from a Baker Perkins cut sheet laminator.

6.2 FAT/FLOUR SPREADING

For some products such as cream crackers, a mixture of fat and flour is spread between each laminated dough sheet. This enhances the separation of the dough sheets and produces an open, flaky texture after baking and a richer flavour with more fat in the cracker (Fig. 6.7).

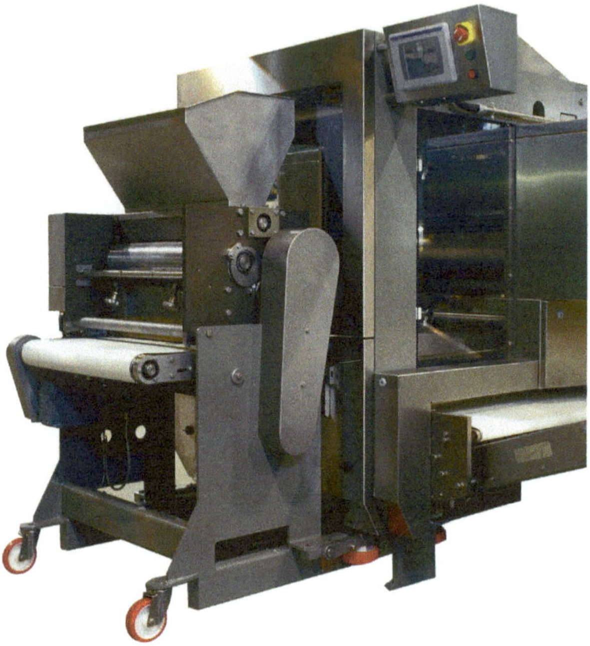

Fig. 6.7 Baker Perkins fat/flour spreader on the left attached to a vertical laminator.

An alternative method of introducing fat/flour for producing cream crackers is to arrange two three-roll sheeters with a fat/flour spreader in between. These sheeters then feed the dough sheets with fat/flour in between into the gauge rolls and layering unit of a vertical laminator.

BIBLIOGRAPHY

Almond, N., 1989. Biscuits, Cookies and Crackers. vol. 2. Elsevier Applied Science.
Baker Perkins Ltd, 2017. www.bakerperkins.com.
Dingson Food Machinery Ltd, 2017. www.dsm-mc.com.
Laser srl., 2017. www.laserbiscuit.com.
Manley, D., 1998. Biscuit Dough Piece Forming, Manual 3. Woodhead Publishing Ltd.
Reading Bakery Systems, 2017. www.readingbakery.com.
Spooner Vicars Bakery Systems, 2017. www.spoonervicarsbakery.com.

CHAPTER 7

Dough Piece Forming: Rotary Moulding

7.1 ROTARY MOULDING PROCESS

Rotary moulding is a versatile forming process for soft, short doughs. A wide variety of shapes, designs and sizes of product are possible. The rotary moulder may be dedicated to one oven, or it may be incorporated into a cutting machine line to provide a multipurpose biscuit line (Fig. 7.1).

Fig. 7.1 Rotary moulded biscuits.

The rotary moulding process consists of pressing the dough into moulds on a roll and then extracting the dough pieces on a cotton canvas extraction web. The moulds have the complete design for the biscuit. There is no scrap as for cut biscuits, and the dough pieces may be transferred directly to the panning conveyor and then to the oven band.

The dough feed to a rotary moulder is important, and the feed must be a metered feed of dough kibbled into small pieces to maintain a constant head of dough in the hopper and an even feed across the width of the hopper. This will assist in ensuring consistent extraction and consistent weights of the dough pieces (Figs 7.2 and 7.3).

Biscuit, Cookie and Cracker Production
https://doi.org/10.1016/B978-0-12-815579-0.00007-6

Fig. 7.2 Rotary moulding dough pieces.

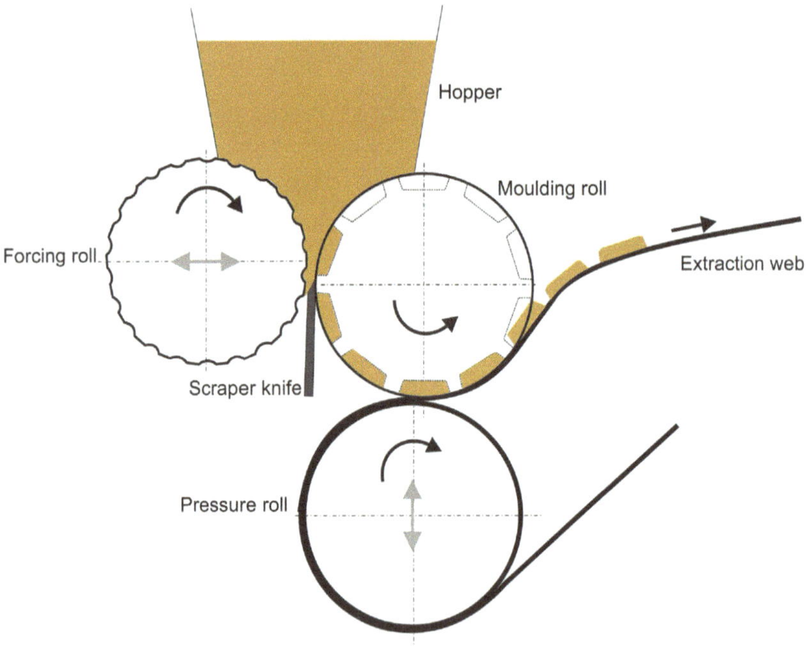

Fig. 7.3 Rotary moulder arrangement.

The forcing roll presses the dough into the moulds, and the dough is sliced off by a scraper knife. Below, the moulding roll is a pressure roll that has a hard rubber surface. This roll supports the extraction web and presses it onto the moulding roll so that the dough pieces will adhere to the web and be extracted from the moulds. The hardness of this rubber surface is critical to good extraction of the dough pieces.

Usually, the gap between the forcing roll and the moulding roll is adjustable, the height of the scraper knife is adjustable, and also, the pressure of the rubber covered the roll below the extraction web. Ideally, there would be three separate drives and speed controls for the forcing roll, the moulding roll and the extraction web to give flexibility for handling different products and recipes (Fig. 7.4).

Fig. 7.4 Baker Perkins Series2 rotary moulder.

7.2 MOULDING ROLLS

The moulding roll may be an engraved bronze roll or a steel roll with plastic inserts for the biscuit impressions (Fig. 7.5).

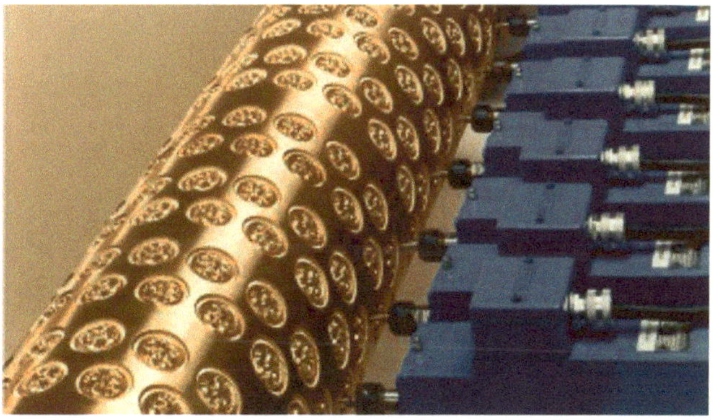

Fig. 7.5 Engraved bronze moulding roll.

BIBLIOGRAPHY

Almond, N., 1989. Biscuits, Cookies and Crackers. vol. 2. Elsevier Applied Science.
Apex Machinery & Equipment Co. Ltd, 2017. www.apexmeco.com.
Baker Perkins Ltd, 2017. www.bakerperkins.com.
Dingson Food Machinery Ltd, 2017. www.dsm-mc.com.
Laser srl., 2017. www.laserbiscuit.com.
Manley, D., 1998. Biscuit Dough Piece Forming, Manual 3. Woodhead Publishing Ltd.
Reading Bakery Systems, 2017. www.readingbakery.com.
Spooner Vicars Bakery Systems, 2017. www.spoonervicarsbakery.com.
Wright, G., 2015. Rotary Moulding. www.biscuitpeople.com.

CHAPTER 8

Dough Piece Forming: Depositing

8.1 DEPOSITING AND WIRE-CUTTING PROCESS

Soft, high-fat doughs for cookies are deposited directly on to the steel oven band. The dough is extruded through dies and then cut by a reciprocating wire. The dough pieces then drop directly on to the oven band (Figs 8.1–8.4).

Fig. 8.1 Deposited cookies.

Fig. 8.2 Cookie depositor over a steel band.

Biscuit, Cookie and Cracker Production
https://doi.org/10.1016/B978-0-12-815579-0.00008-8

Fig. 8.3 Dough pieces being cut and dropped onto the steel band from a wire cut depositor.

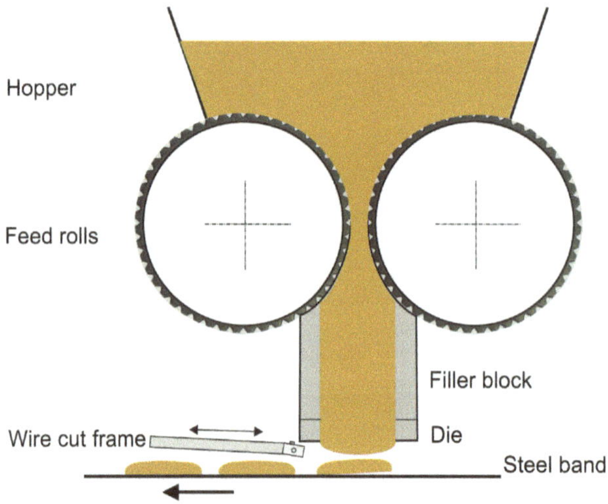

Fig. 8.4 Wire cut machine arrangement.

The soft dough is fed by rolls through a filler block with cylindrical tubes that match the die size. The filler block ensures a smooth passage of the dough from the feed rolls to each individual die, assisting good weight control. A matching filler block and die are used for each cookie. The wire cut frame reciprocates horizontally, and the wire cuts each dough piece as it is extruded through the die (Fig. 8.5).

Fig. 8.5 Baker Perkins: filler block and die withdrawal from a depositor.

8.2 SWIRL TYPE COOKIES

Various alternative products can be deposited without the wire cut mechanism. These include swirl-type cookies, which are extruded through dies, with each die rotating during the depositing process (Figs 8.6–8.8).

Fig. 8.6 Swirl-type deposited cookies.

Fig. 8.7 Swirl cookie depositor with rotating dies from Yesotech Sdn Bhd, Malaysia.

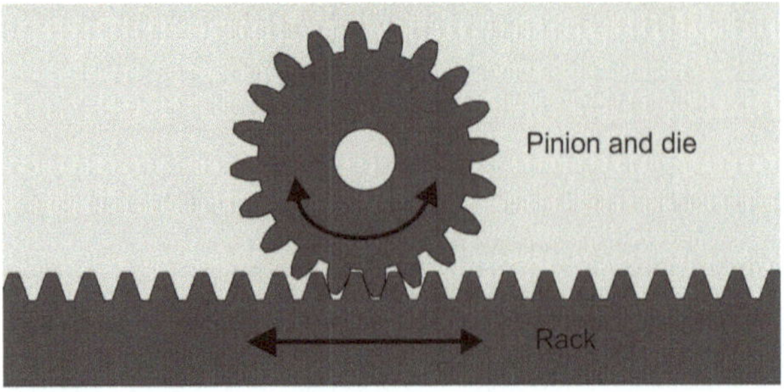

Fig. 8.8 Plan of swirl mechanism: The rack is moved horizontally from side to side causing the pinion and die to rotate.

8.3 FILLED AND TWO DOUGH COOKIES

Deposited products are made by coextrusion of dough and fillings. A further process is rout press, which is a continuous extrusion, usually with a filling. A popular product is fig bar (Figs 8.9–8.11).

Fig. 8.9 Filled cookies.

Fig. 8.10 Fig bars.

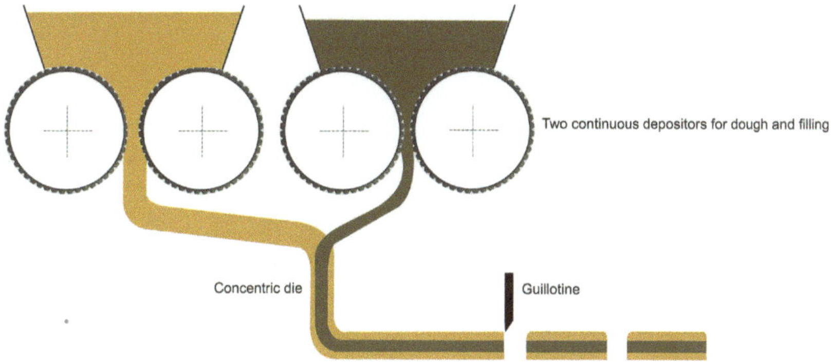

Fig. 8.11 Continuous depositors with concentric dies for rout press products.

8.3.1 Rheon Multi Confectioner and Multi Co Extruder

The Rheon machines are compact, high-performance multirow encrusting machines for confectionery, chocolate cookies, mosaic cookies, marguerite cookies, soft centre, chocolate chip cookies, twisted cookies and cereal bars (Figs 8.12–8.14).

Rheon Multi-Co-Extruder will handle a variety of soft to stiff doughs at high speed, up to 115,200 pieces per hour (16 rows). Options include double filling feeder, wire cut unit and twist nozzle. All extrusion and forming sections are detachable and washable to provide easy product changeover and hygiene.

Fig. 8.12 Rheon product capability.

Fig. 8.13 Rheon multiconfectioner.

Fig. 8.14 Rheon Multi-Co-Extruder.

BIBLIOGRAPHY

Almond, N., 1989. Biscuits, Cookies and Crackers. vol. 2. Elsevier Applied Science.
Baker Perkins Ltd, 2017. www.bakerperkins.com.
GEA Comas SpA, 2017. www.comas.eu.
Manley, D., 1998. Biscuit Dough Piece Forming, Manual 3. 2017. Woodhead Publishing Ltd.
Rheon, 2018. www.rheon.com.
Shanghai Kui Hong Food Machinery Factory, 2017. www.kuihong-foodmachine.com.
Spooner Vicars Bakery Systems, 2017. www.spoonervicarsbakery.com.
Yesotech Sdn Bhd, 2017. www.yesotech.com.my.

CHAPTER 9
Baking Ovens

9.1 BISCUIT BAKING

Biscuit baking ovens, generally known as tunnel ovens, have long conveyors that carry the dough pieces through a heated box section baking chamber. Oven lengths vary typically between 25 and 100 m long. The conveyor band material is a wire-mesh or a carbon steel sheet, which turns around large cylindrical drums at each end of the oven. The conveyor is driven by a variable speed drive at the oven end that allows the operator to adjust the baking time (Fig. 9.1).

Fig. 9.1 Baker Pacific indirect radiant oven, 100 m long.

9.2 HEAT TRANSFER

9.2.1 Radiation

Radiation is the most important method of heat transfer for biscuit baking. It occurs mainly by electromagnetic radiation of infrared wavelengths from direct gas burners, the hot surfaces of the baking chamber and tubes or ducts carrying hot gases from the burners in an indirect baking system. This radiant heat is penetrative and efficient and occurs without adverse side effects, such as the rapid drying or skinning of the surface of the dough pieces.

Biscuit, Cookie and Cracker Production
https://doi.org/10.1016/B978-0-12-815579-0.00009-X

9.2.2 Conduction

Conduction transfers heat from the oven band directly to the base of the dough pieces. The heat transfer is dependent on the temperature and heat mass of the oven band and the surface of the band in contact with the dough piece. With steel bands and heavy mesh (compound balanced weave) bands, this approximates to full contact. Ovens with band preheat can quickly transfer heat into the base of the dough pieces and achieve rapid development of the biscuit structure and texture; this is particularly valuable for cracker baking. Conduction is also important in baking soft cookie doughs on steel bands.

9.2.3 Convection

Convection baking uses hot-air jets that impinge directly on the top surface of the dough pieces and on the underside of the oven band. This system efficiently dries and colours the surface of the dough pieces. However, it produces a dry skin on the dough pieces and will prevent good expansion and 'lift' of the product if used at the start of the baking process.

9.3 OVEN DESIGNS

Ovens are designed to optimise the heat transfer to the dough pieces in different ways. We can group the basic designs into the following:

9.3.1 Radiant Heating

Direct gas fired ovens, electric ovens and indirect radiant ('cyclotherm') ovens (Fig. 9.2).

Fig. 9.2 Baker Pacific direct gas fired/indirect radiant oven.

9.3.2 Conduction Heat Transfer

Ovens with preheated heavy mesh bands and steel bands (Fig. 9.3).

Fig. 9.3 Layer cake baked on a steel band.

9.3.3 'Convection' Baking

Direct and indirect convection ovens transfer the heat by blowing hot-air jets onto the products. Also, 'Recirc' ovens combine convection and radiant heat transfer (Fig. 9.4).

Fig. 9.4 Baker Perkins convection oven.

9.4 DIRECT AND INDIRECT HEATING

The baking chamber may be heated directly with gas burners or electric heaters or by an indirect system using heat exchangers. Direct heating systems use gas or electric energy. The burners or heaters are inside the baking chamber and transfer heat directly by radiation to the dough pieces.

Indirect systems have heat exchangers and may use diesel oil fuel as the products of combustion do not enter the baking chamber. In an indirect radiant oven, the hot air and burnt gas from the burner are circulated through radiant tubes above and below the oven band. These tubes radiate heat to the products. In an indirect convection oven, the burner heats a multipass heat exchanger. Air is circulated through the multipass heat exchanger, and the heated air is blown directly into the baking chamber.

9.5 CONTROL OF THE BAKING PROCESS

The baking chamber of the oven is divided into control zones along the length of the oven. Each zone is typically between 8 and 20 m long and has separate control of the temperature and humidity. The control system allows the baking process to achieve the optimum conditions for the establishment of the biscuit structure, moisture content and colour as the dough pieces travel through the oven.

The control of the humidity in the baking chamber and the removal of moisture from the dough pieces are accomplished by an extraction system in each zone. This consists of ducts that draw air and moisture from the baking chamber through a fan and expel the air through vertical flues (chimneys) to the atmosphere.

In some ovens, this wet air removed from the baking chamber can be diverted either to the flue or back into the baking chamber. This provides moving air within the baking chamber that can aid heat transfer and contribute to even baking conditions across the width of the oven. These systems are called 'turbulence' systems and are mainly used on ovens that have relatively still air in the baking chamber, for example, indirect radiant ovens and direct gas fired ovens.

9.6 DIRECT GAS FIRED OVENS

Direct gas fired ovens are very widely used throughout the biscuit baking industry. They offer versatility to bake all types of biscuits, cookies and crackers (Fig. 9.5).

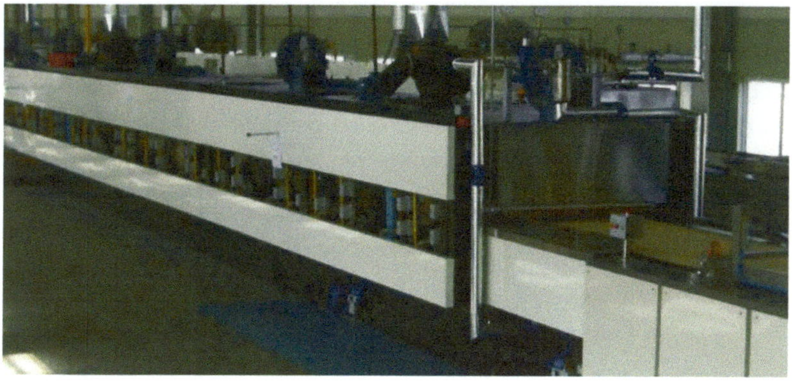

Fig. 9.5 Baker Pacific direct gas fired oven.

The direct gas fired oven has a simple baking chamber of box section with the oven band supported through the middle of the chamber. Above and below the band are ribbon gas burners. A gas/air mixture is supplied to the burner tubes, and this is ignited by a spark electrode and burns on a strip or ribbon across the width of the oven.

The heat transfer in a direct gas fired oven is primarily by radiation from the gas flames and from the oven top, base and walls of the baking chamber (Fig. 9.6).

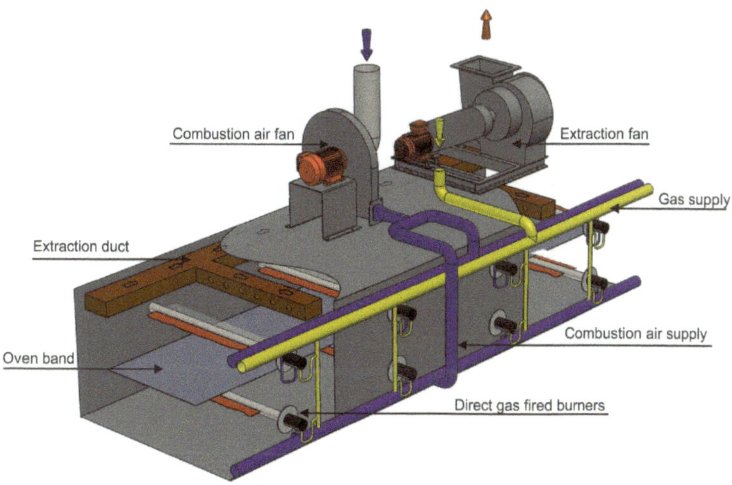

Fig. 9.6 Direct gas fired baking chamber.

Most types of gas including natural gas, town gas (manufactured from coal) and liquid petroleum gas (LPG) may be used.

The burners operate on a 'zero gas pressure' system, and the pressure of the air supplied to the burners is controlled by motorised valves or variable speed blowers to increase or decrease the flame intensity and heat input. Various designs of corrugated stainless steel strips (or ribbons) are used to give a range of heat ratings. Woven wire-mesh (metal fibre) strips are also used for high infrared heat ratings.

This type of oven can achieve high heat inputs per square metre of band surface (up to 40,000 kcal/m^2 may be used for cracker baking) and can successfully utilise any type of baking band, heavy mesh, open wire-mesh bands and steel bands. Direct gas fired burners are used to supply band preheat where this is required for cracker baking (Figs 9.7 and 9.8).

Fig. 9.7 Direct gas fired multifibre burners from Eratec.

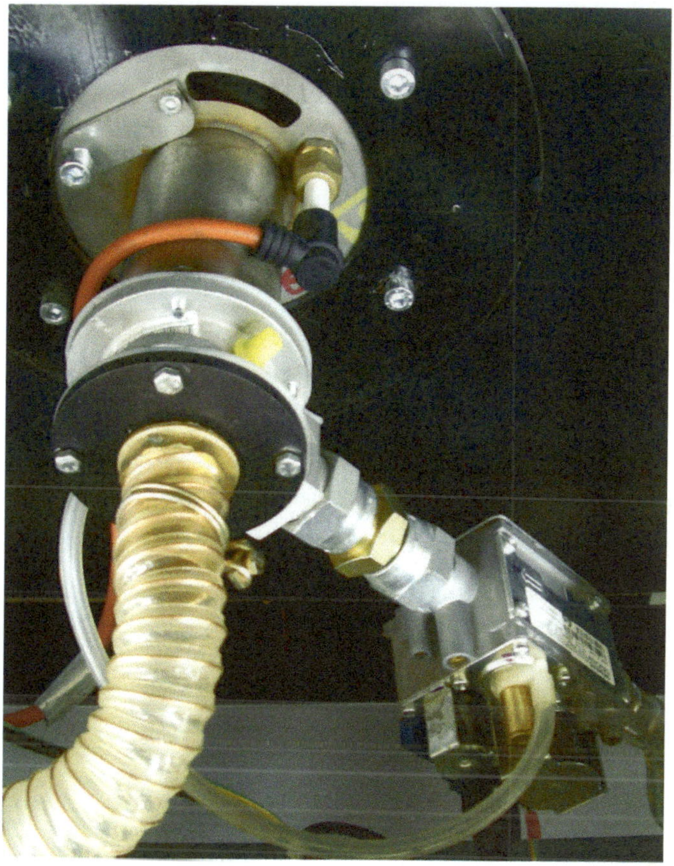

Fig. 9.8 Burner head with gas/air mixer, solenoid valve, ignition and flame detection from Eratec.

Summary

1. Direct gas fired ovens provide radiant heat over a wide range of heat inputs.
2. Rapid and responsive control of heat input and accurate temperature control.
3. Suitable for all types of products: crackers, cookies and all types of biscuits.

9.7 ELECTRIC OVENS

Electric ovens are constructed in a similar way to direct gas fired ovens but use electric heating elements in place of the gas burners. These ovens have been widely used in the baking industry in some countries where industry had adequate electricity supply but lacked gas, for example, China. However, most countries, including China, now use gas predominantly in the baking industry, which is invariably substantially cheaper than electricity.

Summary

1. Electrical energy is clean, does not require storage and is easily controlled.
2. Electric ovens can be simply controlled by thyristor units that vary the current in the elements and hence the heat input to the baking chamber.
3. Suitable for all types of crackers, cookies and biscuits.
4. Very expensive to operate in most markets and not widely used now.

9.8 INDIRECT RADIANT OVENS

The indirect radiant ovens (also known as 'cyclotherm' ovens) are constructed in separate zones. Each zone is typically 8–20 m long, and it has a single burner, heat exchanger and circulation system for the hot gases from the burner (Fig. 9.9).

Fig. 9.9 Baker Pacific indirect radiant oven.

Each zone has a burner firing into a burner tube. The hot burnt gases are drawn from the burner tube through ducts to rows of steel tubes, or ducts, at the top and bottom of the baking chamber. These radiant tubes, or ducts, run the whole length of the zone. The hot gases travel through the tubes or ducts that then radiate heat to the products from above and below. At the end of the zone, the hot gases are collected in a return duct through which they travel back to the circulating fan and from there to the burner tube to be recirculated. It is essentially a closed, circulating system with a single burner, circulating fan and radiant tubes to heat the products from above and below (Fig. 9.10).

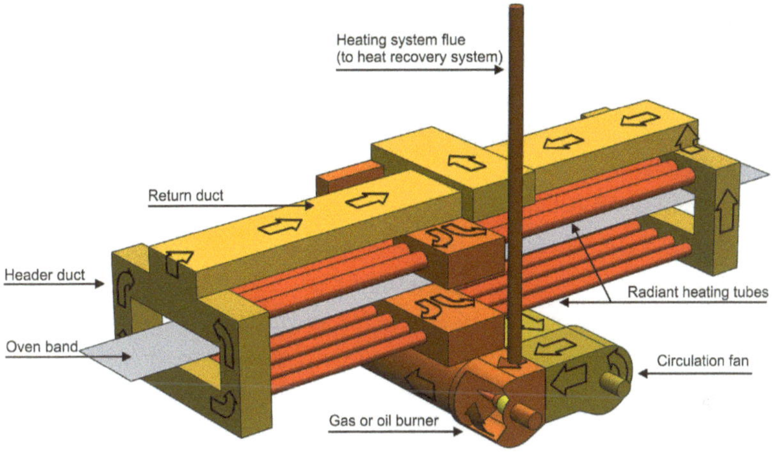

Fig. 9.10 Indirect fired radiant heating system.

A flue with natural convection is used to balance the pressure in the system resulting from the ingress of combustion air to the burner. The continuous recirculation of the hot gases ensures a good efficiency. Fresh air is only drawn into the system at the burner for combustion, and this is balanced by the natural extraction through the burner flue.

Since the products of combustion do not enter the baking chamber, the burner may use diesel oil or gas. This system is commonly used where oil is the most economic fuel, for example, in India.

Summary

1. Indirect radiant ovens provide a very stable, radiant heat, preferred by many bakers.
2. The 'closed' circulation system retains and recirculates the hot gases from the burner and contributes to good fuel efficiency.
3. Suitable for all types of products, except for the first zone(s) of ovens for baking high-rate crackers (e.g. soda crackers) that required direct heating.
4. Suitable for all types of fuel including diesel oil.

9.9 CONVECTION BAKING

Convection ovens are constructed in zones, each zone having a single burner and circulation fan. The fan blows the air around the burner tube, where it is heated, and then through ducts along the length of the zone. These ducts, located above and below the baking band, have slots or nozzles through which jets of hot air are blown onto the products and the oven band. Hot air from the baking chamber is drawn back to the fan to be recirculated through the system. Each zone has an extraction fan and flue to remove moisture from the baking chamber, and this system will also extract the hot air from the heating system.

9.9.1 Direct Convection Ovens

The hot gases from the burner are combined with the recirculated air from the baking chamber and blown through the ducts in the baking chamber and directly onto the products. As the products of combustion are blown directly onto the products, diesel oil fuels are unsuitable for direct convection baking (Fig. 9.11).

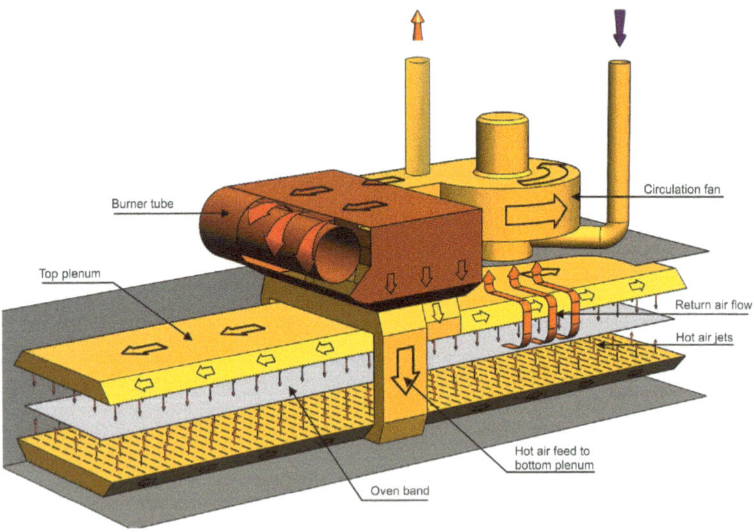

Fig. 9.11 Direct convection baking system (after Haas-Meincke).

9.9.2 Indirect Convection Ovens

The burner fires into a burner tube connected to a multipass heat exchanger. The products of combustion are circulated within the heat exchanger and do not enter the baking chamber. Air is drawn from the baking chamber through the circulating fan and then passed through the heat exchanger, where it picks up heat, before being blown through the ducts in the baking chamber and onto the products. This system can utilise gas or diesel oil fuels (Fig. 9.12).

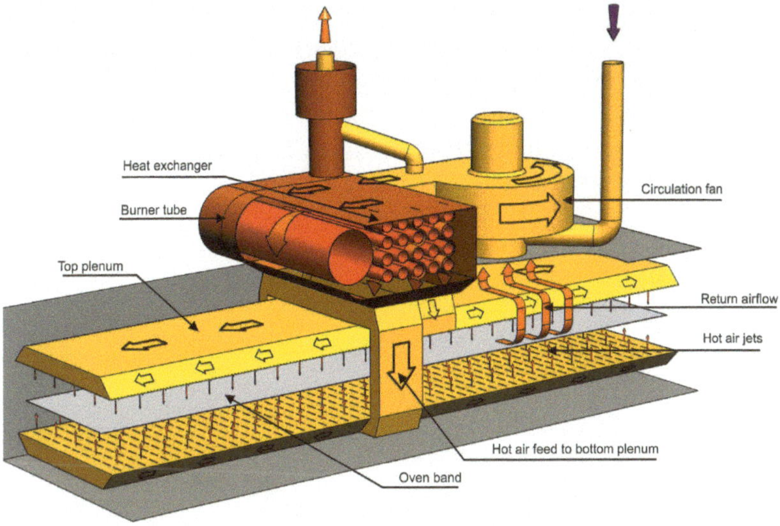

Fig. 9.12 Indirect convection baking system with heat exchanger (after Haas-Meincke).

Summary

1. Convection ovens bake by blowing hot air onto the product surface, so this is not a heat transfer system that optimises the development of biscuit volume and structure.
2. Unsuitable for the first zone(s) of the oven as the convective air dries and skins the surface of the dough pieces preventing good lift and expansion.
3. Effective system for drying dough pieces to low moisture contents. But this rapid drying occurs at the product surface and may contribute to an excessive moisture gradient in the biscuit and consequent 'checking' or cracking of the packaged biscuits.
4. They provide even, bland colouring of the biscuits without contrasts.
5. Direct convection ovens are relatively low-cost ovens to construct and control.

9.10 'RECIRC' OVENS

These ovens were originally developed in the United States as a versatile cookie oven. The system is basically a direct convection oven, but the volume of hot gases blown directly onto the products can be adjusted. When this is reduced, a portion of the hot gases circulate through the ducts at the top and bottom of the oven and return to the burner tube. The system can therefore provide a balance of the heat transfer by convective air and by radiation from the ducts.

'Recirc' ovens may be used for products requiring lower heat inputs. For the first zones, the oven is operated in a mainly radiant mode to avoid 'skinning' the dough piece. However, the heat transfer in this mode is relatively low, and the biscuit structure is slow to form. In the convective mode, the heat transfer is increased, and the system operates as a direct convection oven (Figs 9.13 and 9.14).

Fig. 9.13 Baker Perkins 'Recirc' oven.

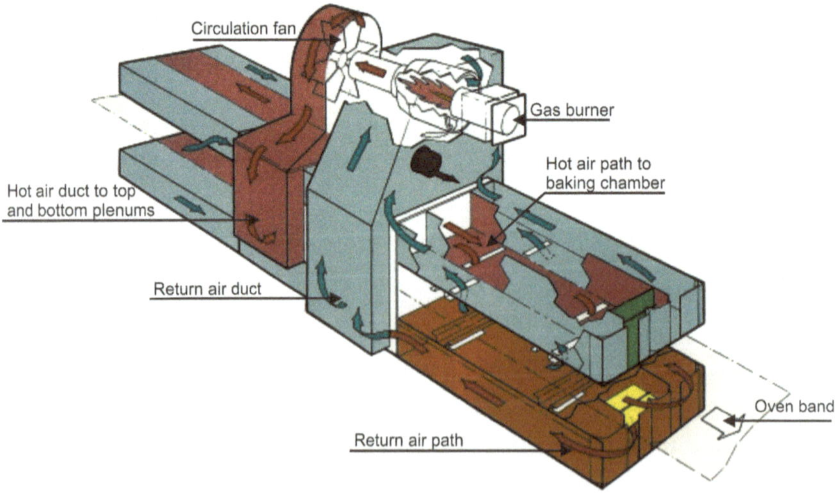

Fig. 9.14 Baker Perkins 'Recirc' oven system.

Summary

1. Suitable for general baking of products requiring relatively low heat inputs
2. Unsuitable for diesel oil fuels

9.11 HYBRID OVENS

It has become common practice to combine different oven types into a 'hybrid' or 'combination' oven. This allows the baker to use different heat transfer modes at different stages of the baking process.

9.11.1 Direct Gas Fired/Indirect Radiant Ovens

Products such as crackers and hard sweet biscuits require high heat inputs in the first part of the baking process to establish good structure and volume. This can only be provided effectively by a direct heating system, and a direct gas fired oven section is normally specified. This system also minimises the drying and skinning of the surface of the dough pieces, which would prevent the lift and expansion of the dough pieces.

The length of this direct fired section is usually one-third of the total length of the oven, and the power input of the direct fired section is one-half of the total power input of the oven.

The indirect radiant oven section will contribute to the optimum development of texture and colour of a wide range of crackers and biscuits (Fig. 9.15).

Fig. 9.15 Baker Pacific hybrid oven: direct gas fired/indirect radiant.

9.11.2 Direct Gas Fired/Convection Ovens

This specification again uses the benefits of the direct gas fired oven in the first third of the oven. The convection section will effectively remove moisture from the dough pieces and achieve a low and even moisture content for the final product. Colour will be even and bland, without contrasts, which is suitable for a range of semi-sweet and short-dough biscuits (Fig. 9.16).

Fig. 9.16 Hybrid direct gas fired/convection oven from Dingson Food Machinery.

9.12 CONDUCTION HEAT TRANSFER

Steel baking bands and heavy mesh bands conduct heat rapidly into the base of the dough pieces. These types of band can be used in any of the oven designs, direct gas fired ovens, indirect radiant ovens and convection ovens.

Steel bands are made of carbon steel, usually 1.2 mm thick. They are principally used for the baking of cookies with high sugar and fat contents, which flow on the oven band in the first part of the oven. Traditionally, steel bands are also used for the baking of 'Marie' (Fig. 9.17).

Heavy mesh baking bands, mainly compound balanced weave (CB) bands, are woven with a tight 'herring bone' pattern providing a solid, thick, heavy mesh. These bands are preheated to 120–180°C, and they conduct

Fig. 9.17 Cookies deposited directly onto a steel band.

heat immediately into the base of the dough pieces as soon as they are deposited on the band. This is a major baking method, being used throughout the industry for the baking of soda crackers and saltines. These bands are also versatile and can be used for a wide range of crackers, semi-sweet and rotary moulded products (Figs 9.18 and 9.19).

Fig. 9.18 Direct gas fired oven with preheat burners.

Fig. 9.19 CB5 compound balanced weave belt.

9.13 BISCUIT TRANSFER FROM THE OVEN BAND

After baking, the biscuits are transferred from the baking band to a wire-mesh stripping conveyor over a knife. The stripping conveyor will include a biscuit reject facility so that any poor-quality product can be rejected by retracting the stripper nosepiece and allowing the biscuits to fall to a cross conveyor for removal to the side of the line.

From the stripping conveyor, the biscuits will be transferred to the cooling conveyors or to an oil spray machine (Figs 9.20 and 9.21).

Fig. 9.20 Biscuits transferred from the baking band over the stripping knife to the wire-mesh stripping conveyor.

Fig. 9.21 Stripping conveyor with retractable nosepiece for biscuit reject.

BIBLIOGRAPHY

Ashworth Bros Inc, 2017. www.ashworth.com.

Baker Pacific Ltd, 2017. www.bakerpacific.net.

Baker Perkins Ltd, 2017. www.bakerperkins.com.

Cambridge Engineered Solutions, 2017. www.cambridge-es.com.

Davidson, I., 1989. A Baker's Guide to Modern Biscuit Ovens. Japanese Biscuit Makers Association.

Davidson, I., 2017. Biscuit Baking Technology, second ed. Academic Press, Elsevier. www.elsevier.com.

Dingson Food Machinery, 2017. www.dsm-mc.com.

Haas-Meincke, 2017. www.haas.com.

Laser S.R.L, 2017. www.laserbiscuit.com.

Reading Bakery Systems, 2017. www.readingbakery.com.

Sandvik Process Systems, 2015. Product Information. www.smt.sandvik.com.

Spooner Vicars Bakery Systems, 2017. www.spoonervicarsbakery.com.

Wire Mesh Products Inc, 2017. www.wire-mesh.com.

CHAPTER 10

Oven Conveyor Bands

10.1 CONVEYOR DESIGN

The products are carried through the oven by a conveyor that has a turning drum at each end of the oven (Fig. 10.1).

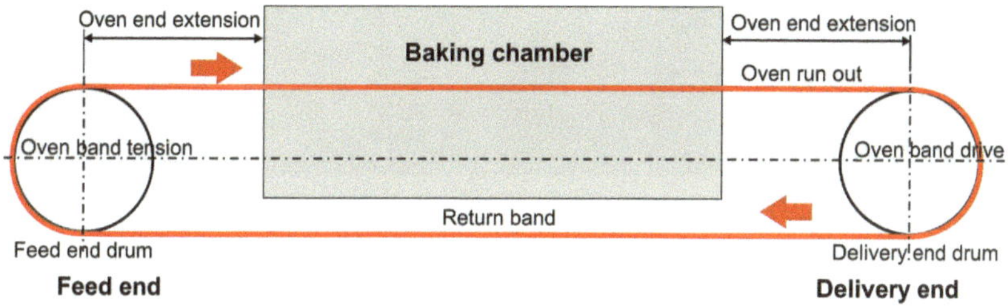

Fig. 10.1 Diagram of oven band circuit.

10.1.1 Feed End

- The feed end unit has the following functions:
- Oven band drum support and movement
- Oven band tension
- Tracking
- Installation of equipment over the oven band (e.g. a wire-cut machine) (Fig. 10.2)

10.1.2 Delivery End

The delivery end unit has the following functions (Fig. 10.3):

- Oven band drum support
- Oven band drive
- Tracking
- Stripping knife and conveyor
- Installation of band cleaner
- Oven end hood and extraction fan

Biscuit, Cookie and Cracker Production
https://doi.org/10.1016/B978-0-12-815579-0.00010-6

Fig. 10.2 Baker Pacific feed end on a cracker oven.

Fig. 10.3 Baker Pacific oven delivery end with oven end hood and fan, transfer to stripping conveyor, band tracker with band position monitor.

10.2 TYPES OF OVEN BAND

The type of oven conveyor band will be selected to suit the product range to be baked. There are three major types of oven bands, each type of band has particular baking characteristics in terms of heat transfer and performance and therefore has a particular range of applications.

10.2.1 Open Wire Mesh Bands

These are the most versatile and widely used oven bands. They are suitable for crackers, semi-sweet biscuits and rotary moulded products, in fact all types of biscuits except soft cookies with high fat contents, where the dough will flow into the holes in the mesh (Fig. 10.4).

Fig. 10.4 Rolled wire-mesh band from Steinhaus GmbH. Copyright Peter Otten of Steinhaus GmbH.

The most popular bands are often referred to as Z47 and Z47R. These designations are from a European supplier, Agrati La Bridoire. Equivalent high-quality wire-mesh bands are available from other suppliers in Europe, the United States and Asia (Fig. 10.5).

Wire Mesh Bands: Skid Bar Supports

Wire-mesh bands are successfully supported in the baking chamber by cast iron skid bars. These bars are placed inside the baking chamber with a pitch of 0.8–1.2 m. They are supported in frames and are removable for cleaning. Apart from cleaning, which is infrequent (normally annually), skid bars do not require any maintenance (Fig. 10.6).

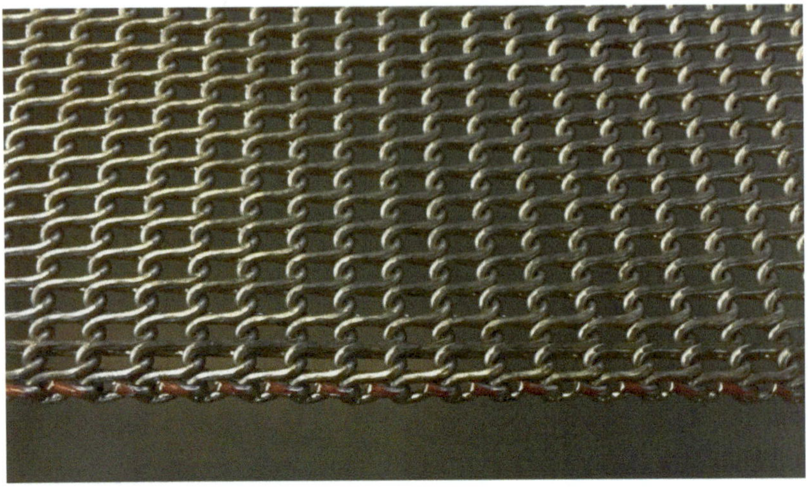

Fig. 10.5 Agrati La Bridoire Z mesh belt with looped edge.

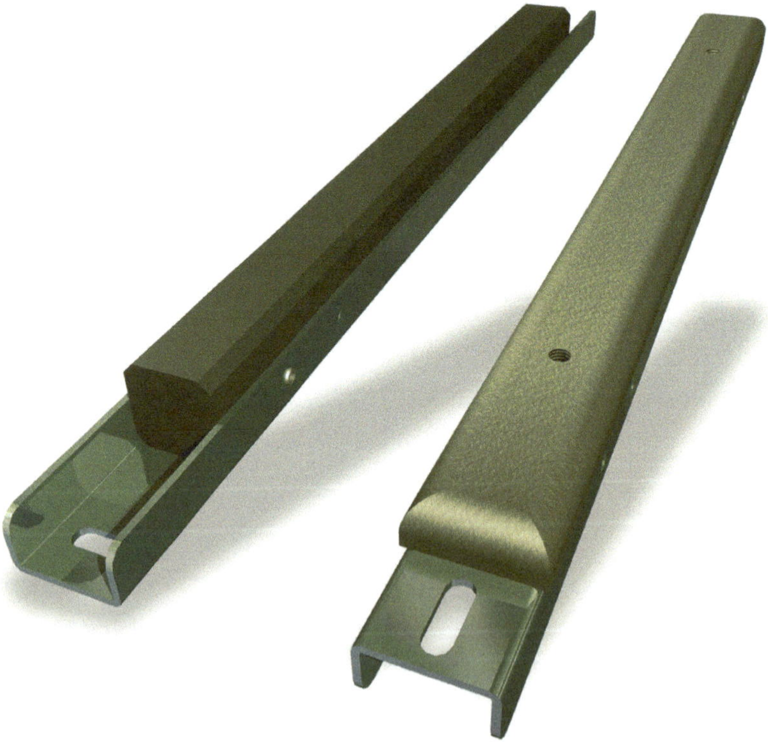

Fig. 10.6 Graphite and cast iron skid bars from Sandvik Process Systems.

The skid bars are cast from iron with a graphite content that reduces friction. The material and casting are critical to the performance.

The installation of the skid bars is critical to good band tracking. The bars must be level and at right angles to the direction of band travel. Levelling is best done optically with a laser instrument.

Return Band Supports

On the return below the baking chamber, the band is supported on rollers. A proportion of these, usually one in three, may have spring loaded guide discs. These are very effective in keeping the band in its central position (Fig. 10.7).

Fig. 10.7 Baker Pacific return band support roller with spring loaded guides.

Wire Mesh Oven Band Cleaning

Oven band cleaning is critical to the quality of the biscuit and the oven performance. We recommend a band cleaner that operates all the time during baking, ensuring that there is no build-up of carbon, which is difficult to remove.

The Baker Pacific oven band cleaner consists of two independently driven wire brushes for cleaning both sides of the band. The unit is designed for a continuous, gentle cleaning action to maintain the band in good condition. The top wire brush, driven by a fixed speed motor gearbox, cleans the inside of the oven band. The lower wire brush unit cleans the outside of the oven band (Fig. 10.8).

Fig. 10.8 Baker Pacific oven band cleaner.

Wire Mesh Oven Band Tracking

It is essential that the oven band runs truly in the centre of the oven in order to avoid damage to the band edges. The position of the oven band edge should be constantly monitored by vertical rollers on each side. The rollers are supported on a bar that follows the movement of the band.

If the band moves to one side by a set distance, an alarm is activated. If the band moves further, the band will stop, and operator action is required immediately to empty the oven of biscuits.

The tracking assemblies are mounted just before each end drum to guide the band onto the drum in a central position. The delivery end tracker is on the top, and the feed end tracker is on the return band at the bottom of the oven. The trackers may have several rolls on a single frame operating together (Fig. 10.9).

Fig. 10.9 Band tracker at delivery end of oven.

Vertical guide rollers are also used inside the baking chamber. Typically, there is one assembly in each oven zone adjacent to the inspection door. These rollers are pivoted and may be spring loaded.

Joining Wire Mesh Bands

To join the ends of a wire-mesh band, two spiral joining wires are used. These are inserted

from the side of the band. After insertion, the spiral joining wires are hammered flat to provide a secure and almost invisible joint (Fig. 10.10).

Fig. 10.10 (1) Inserting a spiral joining wire. (2) The flattened joint is almost invisible. Photos from Steinhaus GmbH. Copyright Peter Otten of Steinhaus GmbH.

10.2.2 Heavy Mesh Bands (Compound Balanced Weave)

These are tightly woven bands with a 'herring bone' pattern presenting a continuous flat surface. They are relatively heavy and have a high heat mass. These bands are preheated and will conduct heat directly into the base of the dough pieces. They are widely used, particularly in North America, for the baking of soda crackers, but are also versatile and may be used for all types of biscuits, except high-fat cookies (Figs 10.11 and 10.12).

Fig. 10.11 Compound balanced weave belt. Rexnord (formerly Cambridge Engineered Solutions).

Fig. 10.12 Compound balanced weave belt. Rexnord (formerly Cambridge Engineered Solutions).

Heavy Mesh Band Supports

Heavy mesh bands are supported on rollers, which will carry heavy bands with low friction.

The design of the supports and adjustment for levelling and the specification of the high-temperature bearings are critical as the rollers operate at baking temperatures. If the rollers seize, they will quickly cause wear to the oven band. The ends of the rollers are therefore exposed for inspection. The bearings are mounted outside the baking chamber and are insulated from the chamber (Fig. 10.13).

Fig. 10.13 Bearing assembly with SKF bearing type Y-V228 suitable for operating temperatures up to 350°C.

Heavy Mesh Band Tracking

Compound balanced weave bands are tracked by vertical side rollers positioned before each drum. These rollers are not intended to force the band into position. They do limit side movement and act as sensing devices to indicate tracking problems.

Joining Heavy Mesh Bands

The bands are joined by inserting the correct number of connecting wires, depending on the type of mesh. The crimped connectors are inserted and seated so that the weave is flat. The connecting wires extend each side by 2 mm and are fastened to the end spirals by a simple weld.

10.2.3 Steel Bands

These bands are made from thin (usually 1.2 mm) carbon steel and are suitable for soft doughs, which flow on the band when heated. They are used for all deposited cookies and cakes, such as layer cake that are baked directly on the oven band. Traditionally, steel bands are also used for baking 'Marie' biscuits (Fig. 10.14).

Fig. 10.14 Cookies deposited directly onto a steel band.

Steel Band Supports

Steel bands are successfully supported on skid bars as described above. These will be a combination of cast iron and graphite skid bars.

Joining Steel Bands

Steel bands may be riveted, and suppliers provide riveting kits including the tools and instructions. However, a much better method is welding, which provides a completely smooth surface to the joint. Welding is a specialist operation carried out by the band supplier with specialist equipment for welding and tempering the joint.

Steel Band Cleaners

Steel bands may be cleaned by rotating brushes (as described above for the wire-mesh bands) assisted by scrapers.

For soft, sticky products such as layer cake, steel scrapers are used. Up to three rows of spring steel scrapers in a staggered formation are mounted on the adjustable frame. For very sticky products, hot water washer units may be used. These are located close to the delivery end of the oven, where the band is still hot.

Steel Band Greasing

At the feed end drum, the band may be greased to prevent the dough sticking. The grease is deposited from a hopper onto soft pads that are mounted on a reciprocating arm. This moves across the band, back and forwards to smear an even coating of grease on the steel band.

Steel Band Tracking

Spring loaded guide rollers provide a robust and safe way to keep a band running within acceptable side movements. A pair of side guide rollers may be installed in each oven zone (Fig. 10.15).

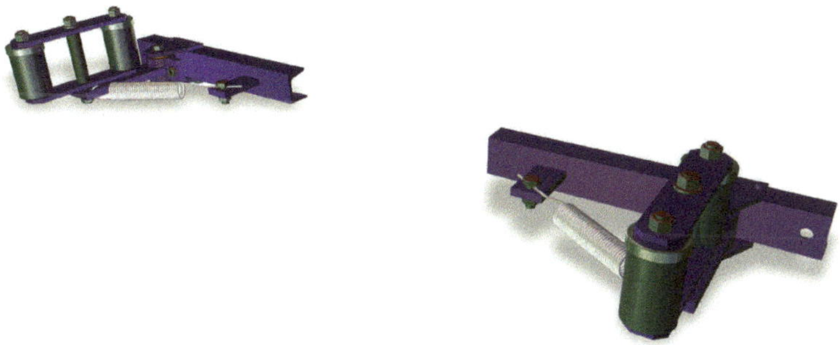

Fig. 10.15 Spring loaded side guide rollers from Sandvik Process Systems.

BIBLIOGRAPHY

Ashworth Bros. Inc, 2014. Balanced Weave Belts. www.ashworth.com.
Baker Pacific Ltd, 2017. www.bakerpacific.com.hk.
Baker Perkins Ltd. n.d. Cleaning wire-mesh bands. www.bakerperkinsgroup.com.
Berndorf Band GmbH, 2014. Bake Oven Belts. www.berndorfband-group.com.
Jan-Ola Jonsson. n.d. The Properties of Steel Belts.
Produits Trefiles de la Bridoire, 2015. Wire bands. www.agrati.com.
Rexnord (formerly Cambridge Engineered Solutions), 2017. www.cambridge-es.com.
Rexnord LLC, 2017. https://www.rexnord.com.
Sandvik Process Systems, 2015. Product Information. www.smt.sandvik.com.
SKF, 2015. Product information. www.skf.com.
Smith, W.H., 1967. Cleaning and Care of Oven Bands. Biscuit Maker and Plant Baker.
Steinhaus GmbH, 2015. Rolled Baking Oven Belts. www.bakingovenbelts.com. www.steinhaus-gmbh.de.

CHAPTER 11

Oil Spray Machines

11.1 OIL SPRAYING

Many crackers are sprayed with vegetable oil immediately after baking. The oil, which may carry flavours, is absorbed quickly when the biscuits are hot. Crackers may have oil contents of 5% up to 17% of their total weight. The oil spray machines may be specified for top or bottom spray or both top and bottom. Special machines can carry flavours as a slurry (Fig. 11.1).

Fig. 11.1 Oil-sprayed crackers.

11.2 OIL SPRAY MACHINES

The oil spray machines carry the biscuits on a stainless steel mesh conveyor through the spray system. The system consists of a pump that supplies the oil to a series of spinning discs that atomise the oil into a fine spay. Multiple discs ensure a consistent oil spray across the width of the machine. Surplus oil drains into a sump and is recirculated to the discs.

It is important that the oil mist is well contained and does not contaminate the adjacent equipment and the floor. A mist filter system is used to contain the fine spray and return it to the sump for recirculation. The oil spray machine conveyor should be at least 2.0 m long and should be followed by a 2.0 m long conditioning conveyor to provide sufficient time to absorb the oil and prevent it from being carried over onto the cooling conveyors (Figs 11.2 and 11.3).

Biscuit, Cookie and Cracker Production
https://doi.org/10.1016/B978-0-12-815579-0.00011-8

Fig. 11.2 Arcall intelli-spin 700 oil spray machine from tna solutions Pty Ltd.

Fig. 11.3 Spinning discs on Arcall 700 oil spray machine from tna solutions Pty Ltd.

BIBLIOGRAPHY

Dingson Food Machinery, 2017. www.dsm-mc.com.
Franz Haas, 2017. www.haas.com.
tna solutions Pty Ltd, 2017. https://www.tnasolutions.com/.

CHAPTER 12
Biscuit Cooling and Handling

12.1 BISCUIT COOLING

Before packaging, biscuits are cooled to near-ambient temperature. This is normally accomplished by a series of conveyors carrying the biscuits for a period of 1.5–2.0 times that of the baking time, when the biscuits will be sufficiently cool for packaging.

This cooling time has a second function, to reduce the moisture gradient from the centre of the biscuits, which is more moist, to the surface of the biscuits, which is very dry. If this moisture gradient is too high, it may cause 'checking' a condition of small hairline cracks in the biscuits after packaging (Fig. 12.1).

Fig. 12.1 Oven delivery end and cooling conveyor.

The cooling conveyors are normally arranged in a two-tier system with three conveyors. The first conveyor is inclined, and the biscuits are then dropped at the end of this conveyor on to a second conveyor running in the reverse direction. The biscuits are then dropped from this conveyor to the third conveyor that will deliver them to the biscuit handling and stacking machines. All the conveyors will have synthetic webs and will be automatically tensioned and tracked (Figs 12.2 and 12.3).

Biscuit, Cookie and Cracker Production
https://doi.org/10.1016/B978-0-12-815579-0.00012-X

Fig. 12.2 Overhead two-tier cooling conveyors with biscuit stacker and packing table.

Fig. 12.3 Biscuit cooling conveyor from AMC System Technology (Suzhou) Co. Ltd., www.coolingtunnelsystem.com.

12.2 BISCUIT HANDLING

After cooling, the biscuits are marshalled for packaging. The biscuit handling system may consist of an automatic lane reduction and stacking facility to channel the biscuits into a suitable arrangement for transfer to the packaging machines.

On single-purpose production lines producing one biscuit, the complete system to feed the packaging machines may be fully automatic. On most production lines, the transfer to the packaging machines will be manual and will accommodate various biscuit sizes and shapes (Fig. 12.4).

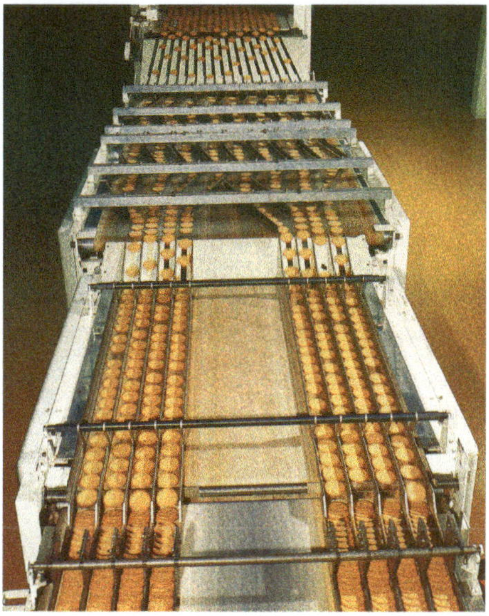

Fig. 12.4 Biscuit channelling board, lane reduction and guides.

The biscuits are stacked in lanes on the packing table for transfer automatically or manually to the packaging machines. The stacking may be effected by a star wheel arrangement or by flip stacking from a fast to a slow conveyor (Figs 12.5–12.9).

Fig. 12.5 Biscuit stacking machine from Dingson Food Machinery, dsm-mc.com.

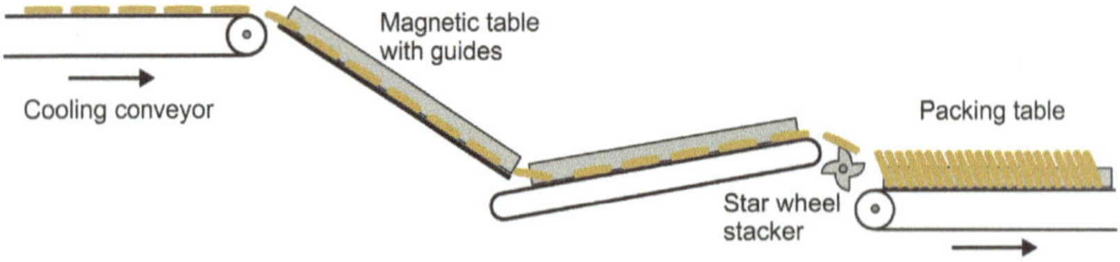

Fig. 12.6 Biscuit stacking arrangement with star wheel.

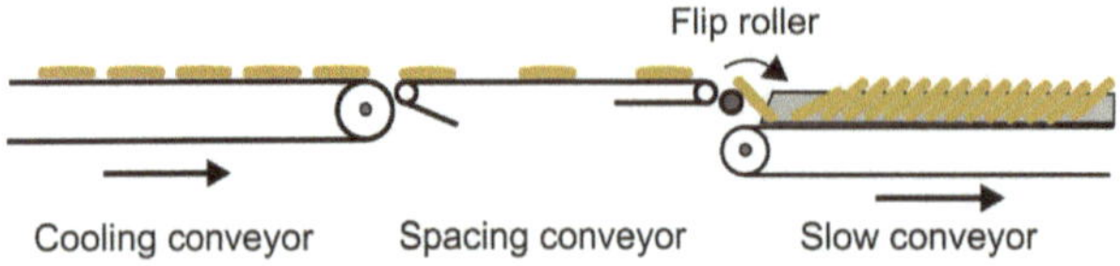

Fig. 12.7 Flip stacking arrangement.

Fig. 12.8 Packing table with manual transfer to trays and packaging machines.

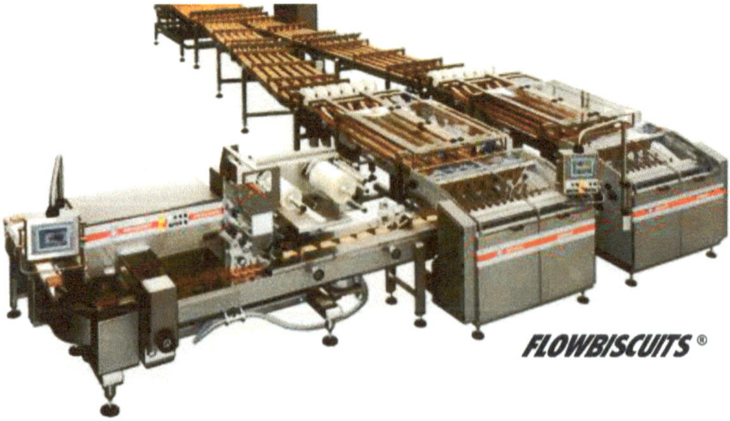

Fig. 12.9 Biscuit handling and wrapping system for biscuits on edge. Receiving, grouping and feeding to flow pack machine from RECORD SPA, www.record.it.

12.3 METAL DETECTION AND CHECK-WEIGHING

Most biscuit lines will have automatic metal detection and check weighing systems for the final packaged products. The systems may incorporate automatic reject systems for the individual packs (Figs 12.10 and 12.11).

Fig. 12.10 Metal detection on a biscuit cooling conveyor from SMMS, India.

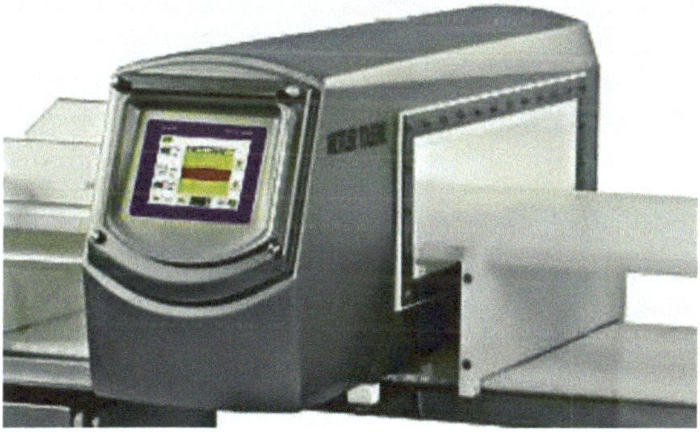

Fig. 12.11 Safeline metal detection for packages from Mettler Toledo.

BIBLIOGRAPHY

AMC System Technology Co. Ltd, www.coolingtunnelsystem.com.
Baker Perkins Ltd, 2017. www.bakerperkins.com.
Dingson Food Machinery, 2017. www.dsm-mc.com.
Houdijk Holland, 2017. https://www.houdijk.com.
Mettler Toledo, 2017. www.mt.com.
Nemesis, 2017. www.checkweighers.eu.
RECORD Spa, 2017. www.record.it.
Riley Product Handling, 2017. www.rileyproducthandling.com.
SMMS, 2017. www.smmsmetaldetector.com.

CHAPTER 13

Biscuit Sandwiching and Chocolate Coating

13.1 BISCUIT SANDWICHING

Biscuits may be automatically sandwiched with cream or jam or two ingredients. The biscuits are normally fed manually from the packing table to a set of vibratory conveyors. The vibratory conveyors ensure a 'pressureless' feed to the sandwiching machine. This avoids breakage of the biscuit shells (Fig. 13.1).

Fig. 13.1 Sandwich biscuits.

13.1.1 Biscuit Magazine Feeds

The biscuit shells are stripped one by one from the feed chutes by pegs on a chain conveyor. This carries the bottom biscuit shell under a stencil that deposits the cream. Subsequently, the top biscuit shell is placed to form the sandwich (Fig. 13.2).

Biscuit, Cookie and Cracker Production
https://doi.org/10.1016/B978-0-12-815579-0.00013-1

Fig. 13.2 Biscuit feed chutes on a Baker Perkins high-speed sandwiching machine.

13.1.2 Cream Feed Systems

The cream is applied by rotary stencils with a wire cut-off. Each stencil is fed by a cream pump from the cream reservoir tank. There is one cream pump for each lane of the sandwiching machine, each with its own variable speed drive and independently adjustable for accurate weight control of the cream deposit. Cream feed speed is set to follow the sandwiching machine (Fig. 13.3).

Fig. 13.3 Baker Perkins cream feed system with two pumps.

13.1.3 Cream Stencils

See Figs. 13.4–13.6.

Fig. 13.4 Baker Perkins cream stencils on a four lane machine.

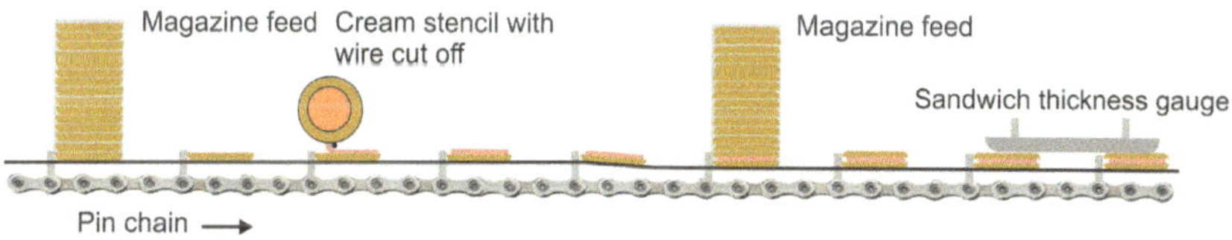

Fig. 13.5 Biscuit sandwiching system.

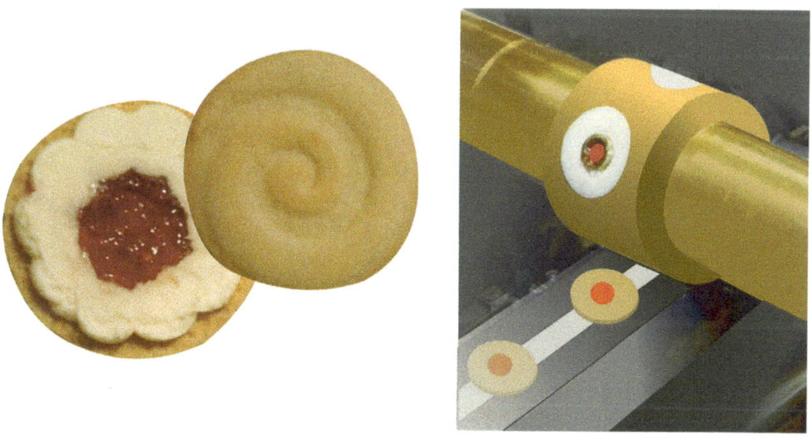

Fig. 13.6 Co-depositing, jam and cream.

13.1.4 Biscuit Cream Sandwich Machines

The machines may have multilanes, usually from two up to six lanes. Each lane may produce 200–800 sandwiches per minute, depending on the product and machine characteristics.

The sandwiching machine will generally be followed by a lane multiplier to increase the number of lanes of sandwiches to feed a cooling tunnel. The cooling will set the cream for the sandwiches to be delivered to packaging (Figs 13.7 and 13.8).

Fig. 13.7 Four lane sandwiching machine with lane multiplier from EverSmart Food Equipment, China.

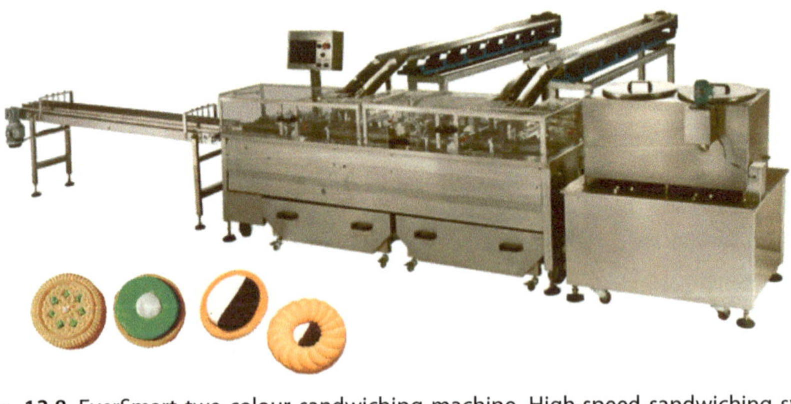

Fig. 13.8 EverSmart two-colour sandwiching machine. High-speed sandwiching system for a variety of biscuit shapes and sizes.

13.1.5 Cookie Capper Machines

Cookie cappers are full width sandwiching systems. The original cookie capper was known as the Mims Capper and is now supplied by Machine Builders & Design.

The cappers take the biscuit shells directly from the end of the cooling conveyor. The machines align the rows of biscuits, flip over the alternate rows, align and deposit the fillings on to alternate rows with bottom side up and then pick and place the top biscuits to form the sandwich.

The systems are versatile and available with different depositing systems to handle cream, jam or mallow. Cappers may have several depositors to enable a variety of fillings to be applied. The cookie cappers may be arranged to feed the sandwiches directly to a chocolate enrobing line (Figs 13.9–13.14).

Fig. 13.9 Korean pies with mallow and cream fillings.

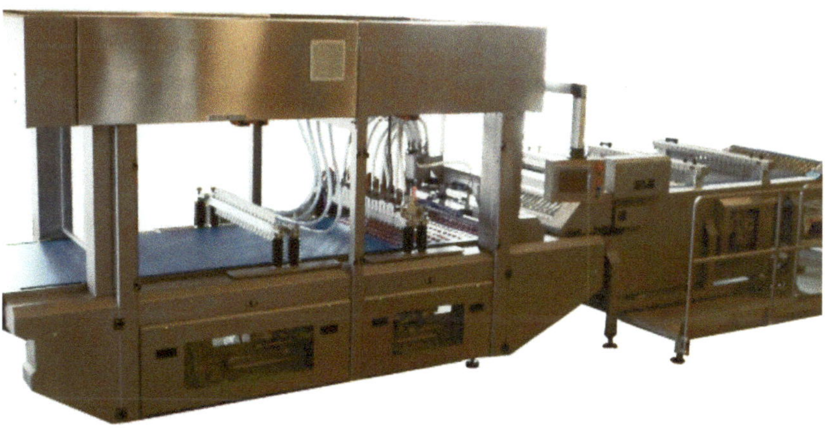

Fig. 13.10 Cookie capper from Procys. www.procys.eu.

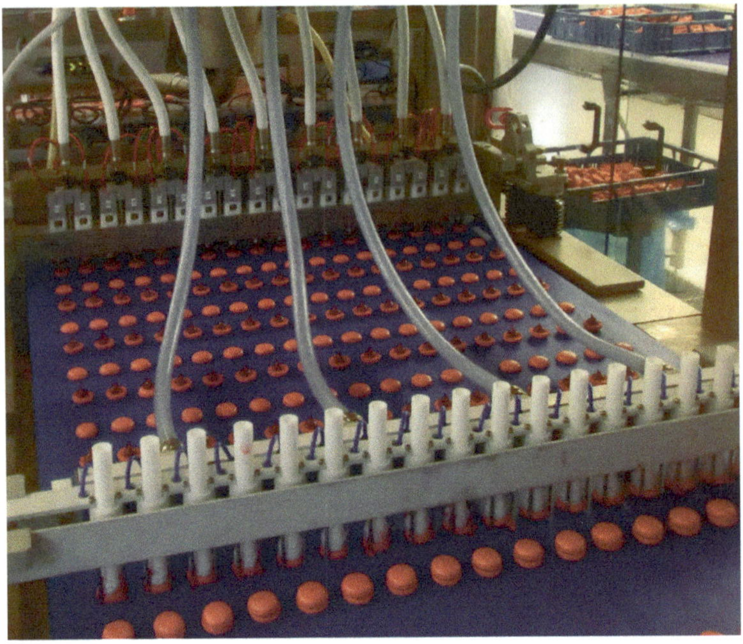

Fig. 13.11 Procys cookie capper.

Fig. 13.12 Procys cookie capper: biscuit alignment.

Fig. 13.13 Procys cookie capper with no product/no cream facility.

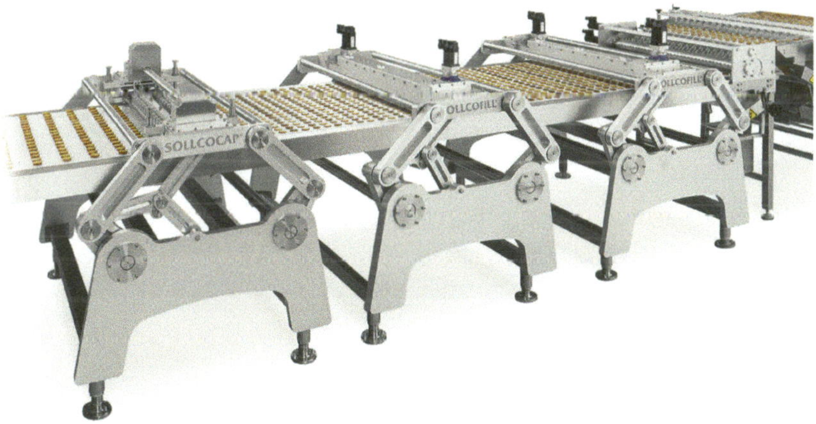

Fig. 13.14 Sollich sandwiching plant Sollocofill®/Sollcocap®.

A recently developed sandwiching system for either single or double deposits with a maximum throughput of up to 240 biscuit rows/min (120 sandwiches per minute).

13.1.6 Sandwich Cooling

Sandwiches are cooled in refrigerated tunnel coolers to set the cream before packaging (Fig. 13.15).

Fig. 13.15 Cooling tunnel from AMC System Technology (Suzhou) Co Ltd. www.coolingtunnelsystem.com.

13.2 CHOCOLATE ENROBING

Chocolate coating of biscuits is a continuous in-line process. The biscuits are delivered directly, after cooling to around 25°C, to an enrobing machine. The biscuits are transported through the enrober on an open wire mesh conveyor and then transferred to a cooling tunnel where the chocolate coating is set (Fig. 13.16).

Fig. 13.16 Chocolate coated biscuits.

Chocolate for enrobing falls into two main categories: pure chocolate with cocoa butter and compound coatings in which the cocoa butter is replaced by a vegetable fat, known as a cocoa butter equivalent (CBE). Cocoa butter melts at a temperature below 37°C, below body temperature, whereas the fats used in compound coatings usually melt at a higher temperature. Compound coatings are less expensive and are more durable in warm climates. Since the coatings on biscuits are generally thin, compound coatings are widely used.

13.2.1 Chocolate Tempering

When pure chocolate is used, accurate temperature control during the process is essential to ensure a glossy coating without fat 'bloom' that appears as a white surface to the coating. The temperature control to seed and produce stable *beta* fat crystals that avoid 'bloom' is carried out in a tempering tube that may be a separate unit or incorporated in the enrobing machine (Fig. 13.17).

Fig. 13.17 Aasted SuperNova tempering unit.

Chocolate is fed from the holding tank to the base of the tempering tube. At the start of the process, the chocolate must be at a temperature of approximately 45°C so that all the crystals are melted. The chocolate is then gently cooled to a temperature of around 24–26°C to initiate the seeding of the stable *beta* crystals. In the final stage of tempering, the chocolate is reheated to 28–29°C before being pumped to the enrober. The temperatures given are a guide, and exact temperatures will depend on the particular chocolate being processed.

13.2.2 Chocolate Enrobing Machines

The biscuits are carried on the open wire mesh conveyor below a curtain of chocolate falling from a flow pan above the conveyor. The chocolate will flow over the biscuits and cover them. Chocolate is also applied to the bottom of the biscuits from a bottoming bath. This facility may be used separately to produce half coated biscuits.

The chocolate is fed from a holding tank via the tempering unit to a tank at the base of the enrober, from where it is pumped to the flow pan and the bottoming bath. Compound coatings may be fed directly from the holding tank to the enrober without tempering.

After coating, the biscuits travel below an air jet, which removes surplus chocolate. There is also a vibrator that shakes the wire mesh band and aids removal of surplus chocolate to leave an even, thin coating before the biscuits are transferred to the cooling conveyor. At the delivery end of the enrober, a small 'anti-tailing' roll is used to prevent tails of chocolate adhering to the edge of the biscuits.

Enrobers may be supplied for full coating and also half coating using a bottoming bath (Figs 13.18–13.20).

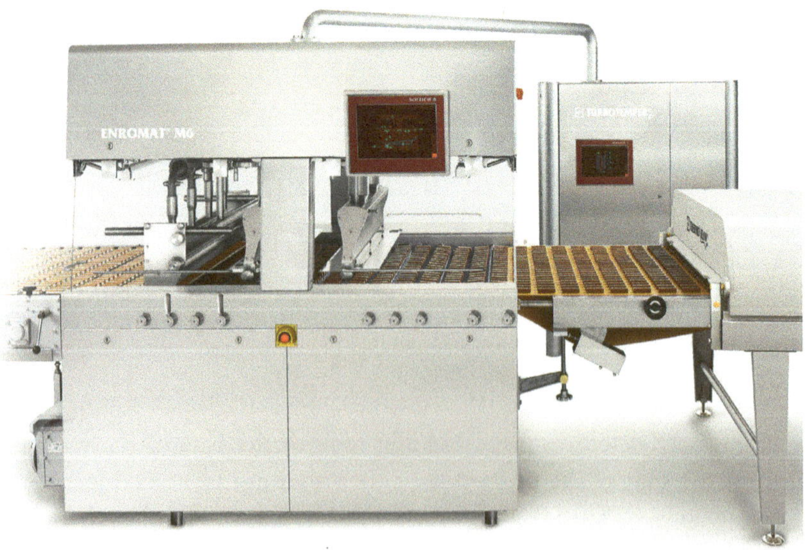

Fig. 13.18 Sollich Enromat® M6 with external tempering unit Turbotemper® type TE.

Fig. 13.19 Chocolate enrober from Dedy showing flow pan (on left) and blower to remove surplus chocolate.

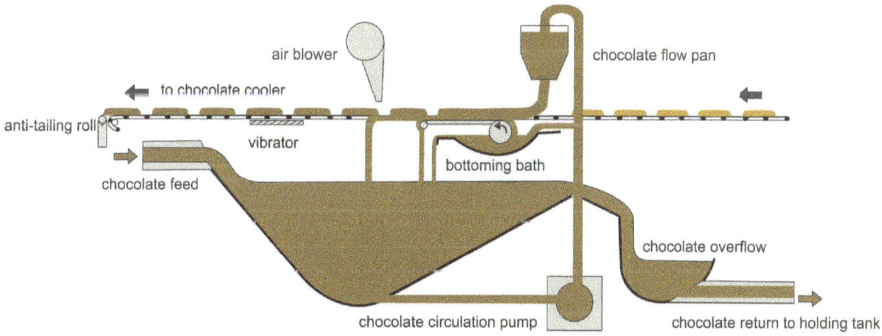

Fig. 13.20 Schematic drawing of enrobing system.

13.2.3 Chocolate Pumps

See Fig. 13.21.

13.2.4 Chocolate Coolers

The biscuits are transported on a synthetic band through a cooling tunnel where cool air is circulated to set the coating prior to transfer to the packaging (Figs 13.22 and 13.23).

Fig. 13.21 Sollich SPN series gear pumps, suitable for feeding tempering machines, coaters and storage tanks; they can feed ring mains and can be used for filling and emptying bulk tankers and conches.

Fig. 13.22 Sollich Enromat® M6 with following cooling tunnel Thermo-Flow Plus.

Fig. 13.23 High-performance chocolate, biscuit and food cooling tunnels for air cooling/water cooling from AMC System Technology (Suzhou) Co Ltd. www.coolingtunnelsystem.com.

BIBLIOGRAPHY

Aasted ApS, 2017. www.aasted.eu.

AAK, 2017.Vegetable oils and CBEs. www.aak.com.

AMC System Technology (Suzhou) Co. Ltd, 2017. www.coolingtunnelsystem.com.

Baker Perkins Ltd, 2017. www.bakerperkins.com.

Beijing Yinrich Technology Developing Co. Ltd, 2017. https://yinrich.en.alibaba.com.

Dedy GmbH, 2017. www.dedy.de.

Eurosicma Spa, 2017. www.eurosicma.it.

EverSmart Food Equipment, 2017. www.eversmartpack.com.

Food Processing Equipment s.r.l, 2017. www.fpe2000.it.

GEA Group, 2017. www.gea.com/en/productgroups/food-processing-systems/bakery-equipment/index.jsp.

K&Kurt Makina, 2017. www.kurtmakina.com.

Peerless Food Equipment, 2017. www.peerlessfood.com.

Procys, 2017. www.procys.eu.

Production Techniques Ltd, 2017. www.ptl.co.nz.

Richeng Foodstuff Machinery Co. Ltd, 2017. www.ri-cheng.com.cn.

Riley Product Handling Ltd, 2017. https://www.rileyproducthandling.com.

Sollich KG, 2017. www.sollich.com.

CHAPTER 14
Biscuit Packaging

14.1 TYPES OF BISCUIT PACKAGING

See Fig. 14.1.

Fig. 14.1 Types of biscuit pack: vertical form fill seal, carton, box, flow pack (fin seal), roll or slug wrap (lap seal) and trays.

14.2 PACKAGING FUNCTIONS

- Present the biscuits in an attractive way for the potential consumers
- Display the biscuit type, weight, ingredients and manufacturer
- Maintain biscuit freshness and flavour over a long shelf life

Biscuit, Cookie and Cracker Production
https://doi.org/10.1016/B978-0-12-815579-0.00014-3

- Provide effective barrier for moisture and foreign odour
- Resist seepage of fats and oils
- Protect against visible light and UV radiation
- Protect the biscuits from damage during transportation
- Stack well on supermarket shelves

14.3 PACKAGING MATERIALS

The packaging materials are selected to provide a moisture barrier and to protect the biscuits. The film materials are heat or cold sealable. Some of the various materials used are the following:

- Polypropylene, metallised orientated polypropylene (OPP) and biaxially oriented polypropylene film (BOPP).
- BOPET, biaxially oriented polyethylene terephthalate (PET), is a polyester film made from stretched PET.
- Laminates, plastic-coated papers, OPP/paper and metallised polylactic acid/paper.
- For thermoformed trays, PET and recycled PET (PET).
- Aluminium foil.
- Cartons and boxes: paperboard, coated paperboard and paperboard/polyethylene/aluminium foil.

14.4 MODIFIED ATMOSPHERE PACKAGING

Modified atmosphere packaging (MAP) is the process of removing oxygen from inside the package and replacing it by a gas flushing with a mix of oxygen and inert gases such as nitrogen and carbon dioxide. The process improves the freshness and shelf life of the biscuits.

14.5 VERTICAL FORM FILL SEAL PACKAGING

Small snack biscuits are normally packed in vertical form fill seal (VFFS) bags or pillow packs. The biscuits are fed in bulk to a multihead weigher, which divides the biscuits into the individual pack weights. The VFFS machine forms the packs from a flat roll of film and simultaneously fills the bags with a preweighed quantity of biscuits (Figs 14.2–14.4).

The Bosch SVI 2610 is ideal for midspeed, standard bag styles, such as pillow, gusseted and block bottom bags. It can package a wide range of products including biscuits, snacks, grains and confectionery (Figs 14.5–14.8).

Fig. 14.2 Mini biscuit pack.

Fig. 14.3 Fragile product Multihead Weigher CCW from Ishida Co. Ltd.

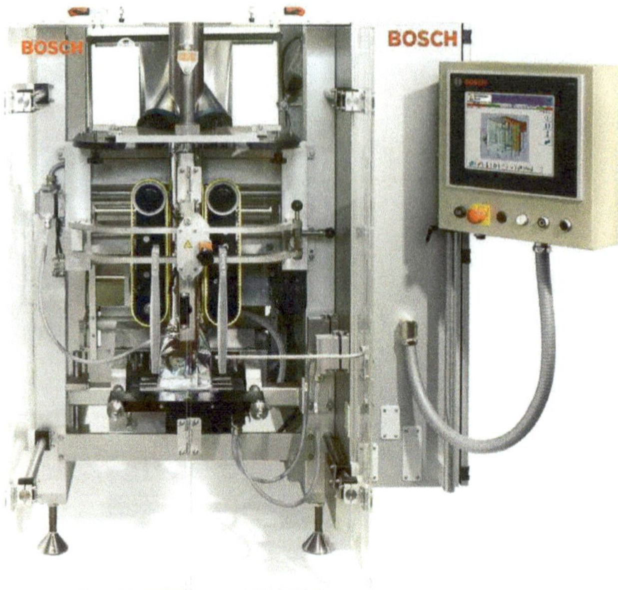

Fig. 14.4 Bosch SVI 2610 vertical form fill seal machine (picture, Bosch).

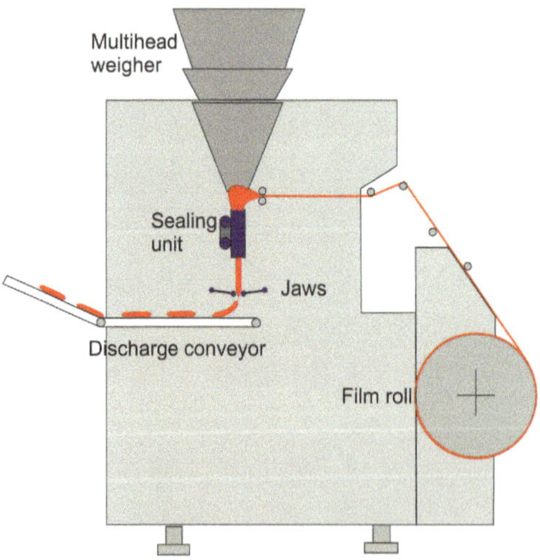

Fig. 14.5 Schematic drawing of vertical form fill seal system.

Fig. 14.6 Snack pack.

Fig. 14.7 Mini biscuits bags.

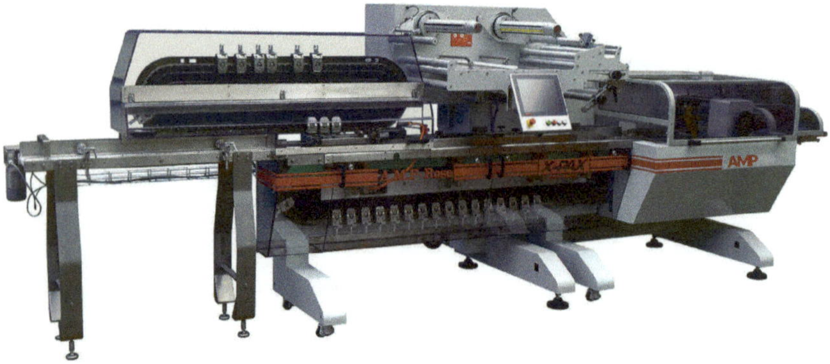

Fig. 14.8 The X-Pax flexible multipack wrapping machine is a horizontal form fill and seal machine, which has been specifically designed for multipacking items such as biscuit products. A.M.P. Rose. amp-rose.com.

14.6 ON EDGE FLOWPACK

The infeed chain of a horizontal flow pack machine is fed by groups of biscuits. The film is fed from a reel above and folded around the biscuit groups as a tube. The film is heat sealed below as a fin by rotating sealers. The biscuit groups within the film tube then transfer in line to the crimping, sealing and cutting section of the machine (Figs 14.9–14.13).

Fig. 14.9 Cavanna Zero 5 NKZ wrapping machine for slug products with automatic feed. www.cavanna.com.

Fig. 14.10 On edge flow pack.

Fig. 14.11 Pack 301 LS (length slug) flow wrapper (picture, Bosch).

Fig. 14.12 Biscuit slug feed (picture, Bosch).

Fig. 14.13 On edge flow pack slug packs.

14.7 BISCUITS IN TRAYS

See Fig. 14.14.

Fig. 14.14 Flow pack packaging of biscuits in thermoformed plastic trays.

14.8 PILE PACKS

A continuous loader for horizontal flow pack machines can form biscuits in piles of two and pack one or two piles of biscuits from a single lane in feed. The system can handle up to 400 biscuits per minute (Figs 14.15–14.19).

Fig. 14.15 Pile pack.

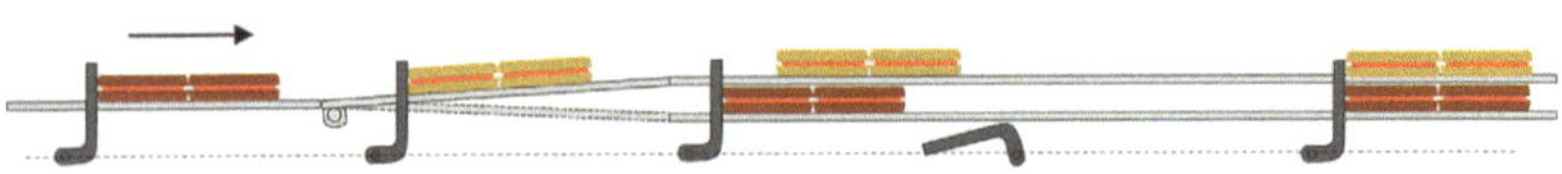

Fig. 14.16 Pile pack forming system.

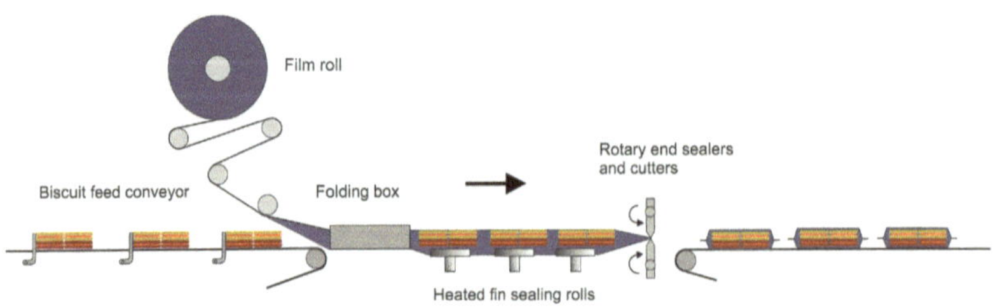

Fig. 14.17 Flow pack system for pile packs.

Fig. 14.18 Flow packs.

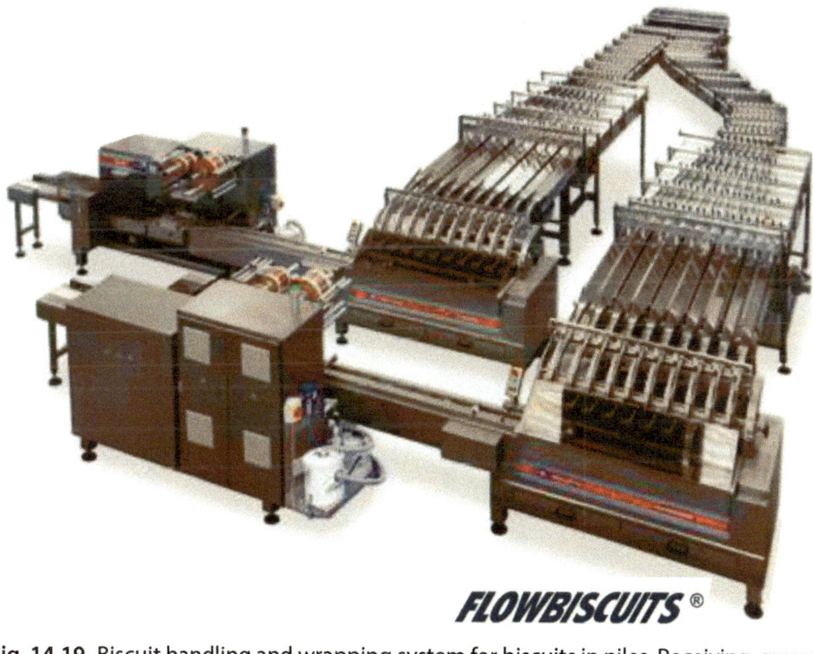

Fig. 14.19 Biscuit handling and wrapping system for biscuits in piles. Receiving, grouping and feeding to flow pack machine from RECORD SPA. www.record.it.

14.9 CARTONS

Direct feed with lock-type forming or indirect feed with glue-type forming. Single- or double-forming head. Cama Forming machines are suitable to form single- or triple-flap boxes (Figs 14.20–14.22).

Fig. 14.20 Carton packaging for biscuits.

Fig. 14.21 Carton packaging for snack biscuit packs.

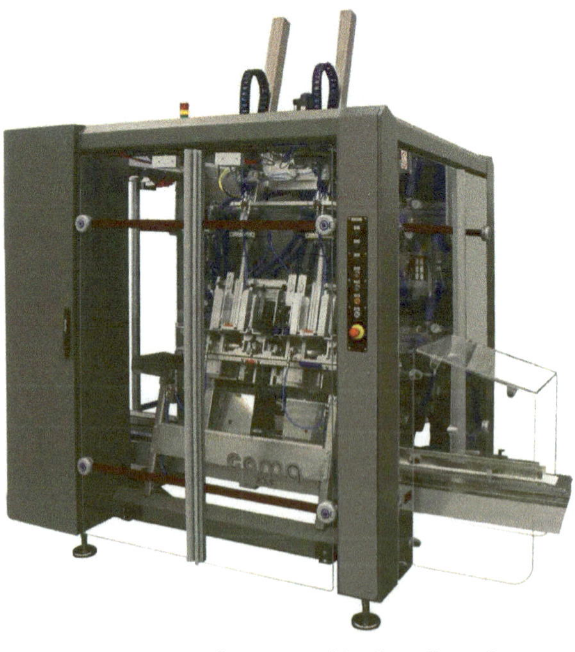

Fig. 14.22 Lock- and glue-type carton forming machine from Cama Group, www.camagroup.com. Tray and box electronic forming machines from cardboard, 'e-flute' or corrugated blanks. Direct feed with lock-type forming or indirect feed with glue-type forming. Single or double forming head. Cama Forming machines are suitable to form single or triple flaps boxes.

14.9.1 Top Loading Cartons

See Figs. 14.23–14.28.

Fig. 14.23 Top-loaded carton.

Fig. 14.24 Top-loaded carton with opening flap.

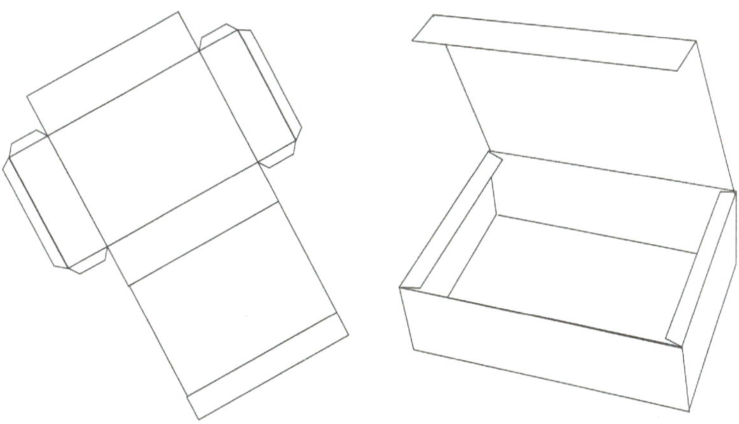

Fig. 14.25 Top-loading carton forming.

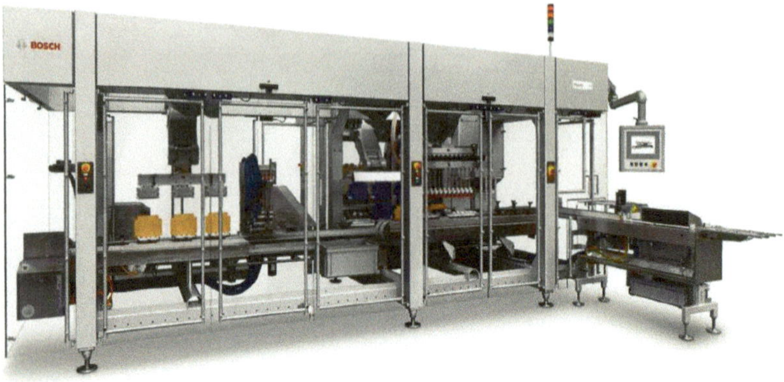

Fig. 14.26 Sigpack TTM integrated top-loading cartoner for up to 100 cartons per minute (picture, Bosch).

Fig. 14.27 Kliklok integrated top-load cartoner (picture, Bosch).

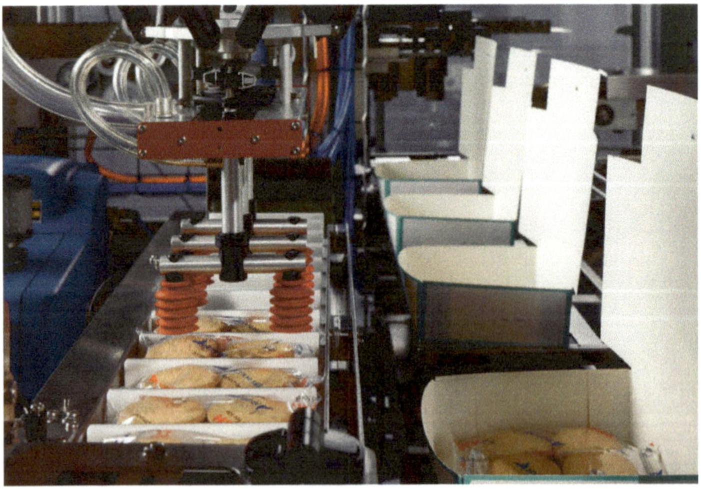

Fig. 14.28 Kliklok ITC machine (picture, Bosch).

14.9.2 Vertical Cartons

See Figs. 14.29–14.31.

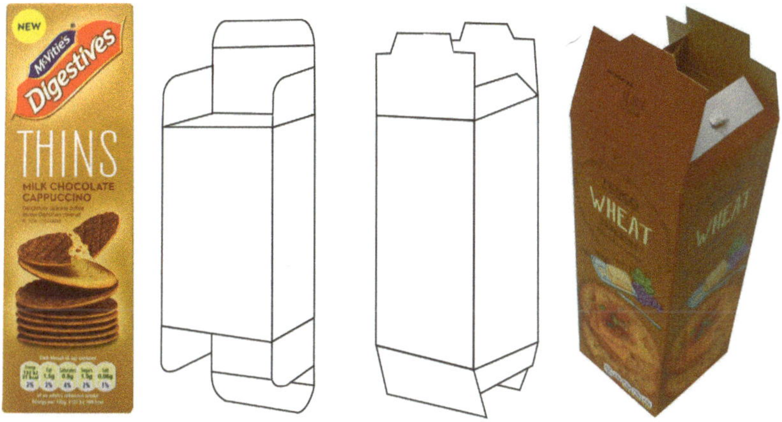

Fig. 14.29 Vertical cartons.

Fig. 14.30 Bag-in-box packaging.

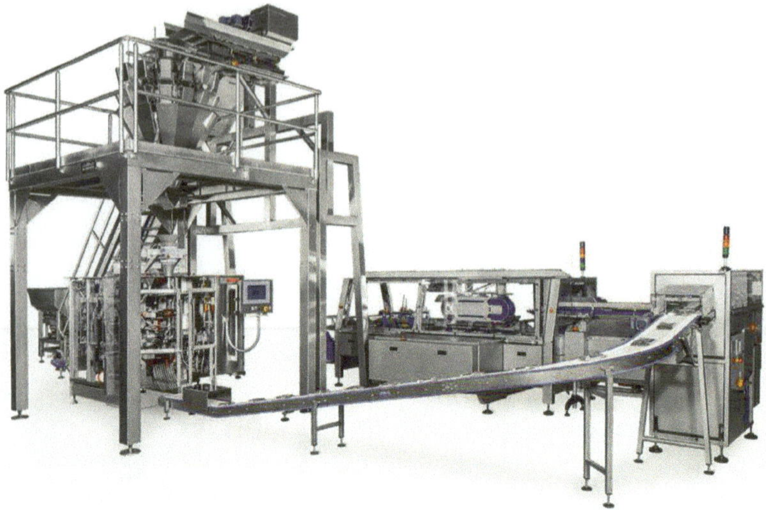

Fig. 14.31 Bag-in-box system (picture, Bosch).

14.9.3 Horizontal Cartoners

Cama Group provides electronic intermittent-continuous motion cartoning machines for packing flow wraps, pouches, bags, thermoformed trays, blisters jars and more. The machine functions based on loading and closing two or more cartons with each machine pitch by means of an intermittent-continuous motion. Carton closing by hot melt or tuck-in (Figs 14.32–14.34).

Fig. 14.32 Horizontal carton.

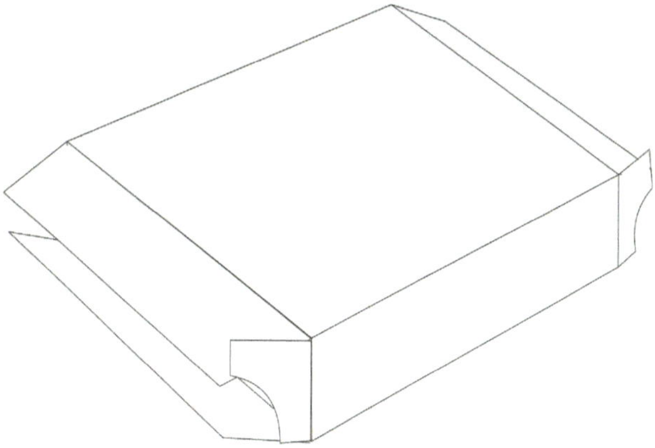

Fig. 14.33 Horizontal carton form.

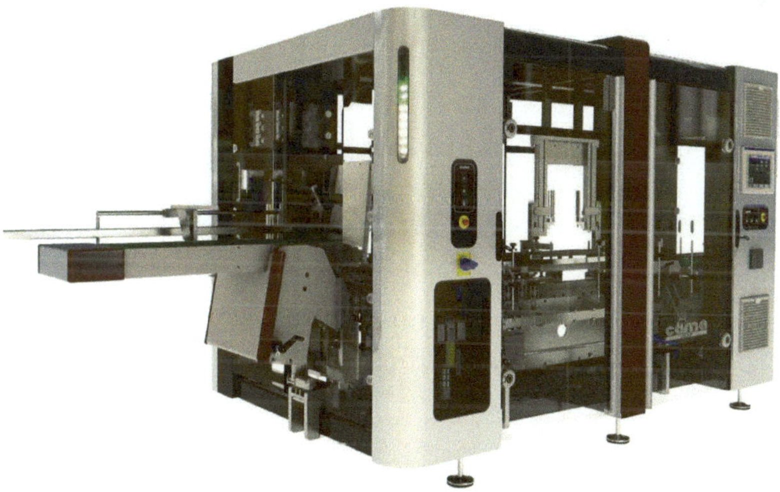

Fig. 14.34 Side loading horizontal cartoner from Cama Group. www.camagroup.com.

14.10 BISCUIT TINS

Biscuits in tins are widely appreciated as gifts (Fig. 14.35).

Many packaging tasks, previously always done manually, may now use 'pick and place' automatic systems and robots. Schur Technology's 'Tray-, Box-, Tin Automatic F350' is designed for fully automatic packing of pleated paper cups containing cookies. The system is used by leading manufacturers of Danish butter cookies (Figs 14.36–14.38).

Fig. 14.35 Danish butter cookies from PT Mayora Indah.

Fig. 14.36 Schur F350 packaging machine for Danish butter cookies. https://www.schur.com.

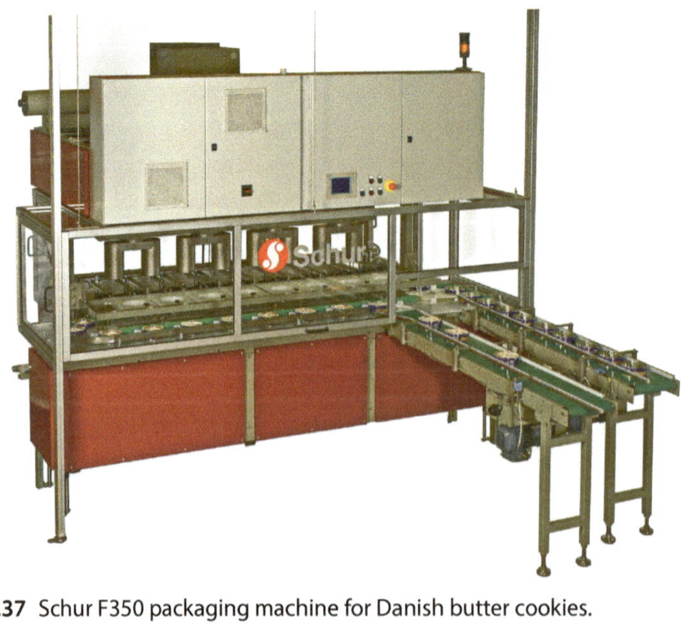

Fig. 14.37 Schur F350 packaging machine for Danish butter cookies.

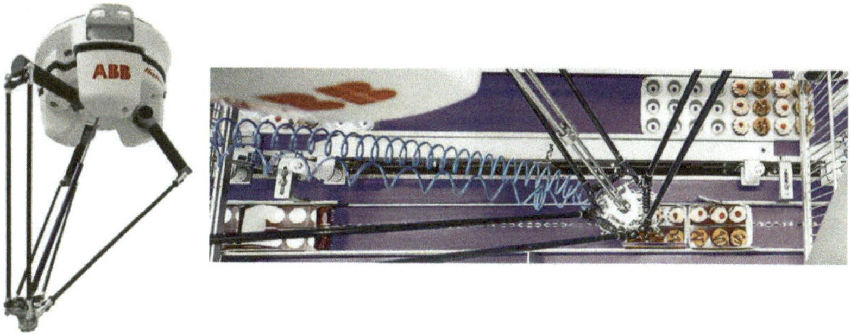

Fig. 14.38 ABB IRB 360 robotic arm picking and placing bakery items. © ABB.

BIBLIOGRAPHY

ABB Ltd, 2017. www.new.abb.com/products/robotics.
A.M.P. Rose, 2017. www.amp-rose.com.
Anton Ohlert, 2017. www.ohlert.com.
Apex Machinery & Equipment Co. Ltd, 2017. www.apexmeco.com.
ATP Engineering & Packaging S.L, 2017. www.atp-packaging.com.
Avoncourt Packaging Ltd, 2017. www.avoncourt.com.
Bosch, R., 2017. www.boschpackaging.com.
Bradman-Lake, 2017. www.bradmanlake.com.
Cama Group, 2017. www.camagroup.com.
Cavanna Packaging Group, 2017. www.cavanna.com.
Eurosicma Spa, 2017. www.eurosicma.it.
Fuji Machinery Co. Ltd, 2017. www.fuji-machinery.com. www.fuji-packaging.com.
Systems, H.F., 2017. www.hayssen.com.
Hayssen Flexible Systems, 2017. www.hayssen.com.
Holland, H., 2017. www.houdijk.com.
Houdijk Holland, 2017. www.houdijk.com.
Ilapack, 2017. www.ilapack.com.
Innovia Films, 2017. www.innoviafilms.com.
Ishida Co, 2017. www.ishida.com.
Jornen Machinery Co. Ltd, 2017. www.jornen.com.
Kawashima Packaging Machinery Ltd, 2017. www.kawashima-pack.co.jp.
Khosla Precisions, 2017. www.biscuit-packaging.com.
Leeways Packaging Services Ltd, 2017. www.leeways.co.uk.
Multihead Weighers, 2017. www.multiheadweighers.co.uk.
O-M Ltd, 2017. www.ns.0.1.omltd.co.jp.
Qingdao Nissin Food Machinery Co. Ltd, 2017. www.meetnissin.com. www.nissinpacking.en.made-in-china.com.
Record Packaging Machines, 2017. www.record.it.
Reedbut Group Ltd, 2017. www.reedbut.com.
Schib, 2017. www.schib.it.
Schur Packaging Systems AB, 2017. https://www.schur.com.
Serpa Packaging Solutions, 2017. www.serpapackaging.com.
Ulma, 2017. www.ulmapackaging.com.
Universal-Pack, 2017. www.universalpack.it.

CHAPTER 15

Biscuit Production

The production of each type of biscuits, cookies and crackers requires a different range of ingredients, formulations and production equipment. The production process is described for examples of the main biscuit categories. For information on the biscuit ingredients, please see Chapter 16.

15.1 PRODUCTION OF SNACK CRACKERS

See Fig. 15.1.

Fig. 15.1 Snack crackers.

15.1.1 Description

Snack crackers are successful in every market: light and crispy with oil spray.

Product Specification

Dimensions	48.0 mm diameter
Thickness	4.9 mm
Weight	3.0 g

Biscuit, Cookie and Cracker Production
https://doi.org/10.1016/B978-0-12-815579-0.00015-5

Appearance	Evenly blistered
Colour	Golden
Texture	Light and crispy
pH	5.5
Moisture	1.3%–2.5%

Formulation

Flour	100.00
Sugar	8.02
High-fructose corn syrup	2.85
Vegetable oil (soya bean)	11.66
Lecithin	0.20
Ammonium bicarbonate	1.84
Sodium bicarbonate	1.08
ACP	1.08
Salt	0.77
Enzyme	0.01
Water	29.47

15.1.2 Mixing

An 'all in' mix on a horizontal mixer. All the ingredients are fed into the mixing bowl at the start of the mixing process. The mixing is continued until a temperature of typically 33°C is reached for enzyme doughs (Fig. 15.2).

Fig. 15.2 Baker Perkins horizontal high-speed mixer with shaft-less blade.

Standing Time

After mixing, the dough is stood to allow the enzyme to react with the gluten. The standing time is about 2.0–2.5 h at 35°C. The time must be determined carefully depending on the amount of enzyme and the quality of the flour in order to achieve the soft, delicate eating texture of the cracker.

15.1.3 Forming

The dough is laminated with four laminations, approximately 4 mm thick. Fat/flour filling is not used (Fig. 15.3).

Fig. 15.3 Baker Perkins forming line: (from right to left) laminator, three gauge roll units, relaxation conveyor, rotary cutter, scrap lift and return conveyors.

15.1.4 Baking

Snack crackers may be baked on a direct gas-fired oven or a hybrid oven (Fig. 15.4).

Fig. 15.4 Baker Pacific hybrid oven for snack cracker production.

Baking time	5.0 min
Zone temperatures	220/220/230/230/180°C
Band preheated to	150–180°C

High bottom heat in zones 1 and 2 of the oven is required.

Baking Band

Compound balanced weave or Z47 type wire-mesh band.

Preheat is required.

15.2 PRODUCTION OF SODA CRACKERS

See Fig. 15.5.

Fig. 15.5 Soda crackers.

15.2.1 Description

Soda crackers are a traditional product in the United States, where they are made in very large volumes.

Similar crackers are 'saltine' or 'premium' crackers. The process has several important characteristics:

- A two-stage mixing process known as 'sponge and dough'.
- A long fermentation, usually 24 h.
- Fast baking time, around 2.5 min, on a heavy mesh preheated oven band.
- Soda crackers are baked in strips or sheets and are broken into individual crackers, or pairs after baking.

Product Specification

Dimensions	91 × 44 mm
Thickness	5.6 mm
Weight	6.25 g
Appearance	Evenly blistered
Colour	Pale creamy colour with darker blisters, evenly spaced
Texture	Open and flaky, with a crispy bite
Flavour	Mild, fermented flavour
pH	7.2–8.0
Moisture	2.5%

Formulation

Sponge

Flour (strong)	66.7
Fresh yeast	0.17
Dough fat	5.00
Lecithin	0.53
Malt extract 80%	0.95
Water	28.0

Dough

Sponge (as above)

Flour (weak)	33.3
Dough fat	5.00
Soda	0.60
Salt	1.50

Critical Ingredients

A strong flour produced with 30% of hard wheat will give an open cracker texture. The flour used in the sponge must be 10%–11% protein. Stronger flour gives a harder cracker.

A weaker flour (8.0%–9.0% protein) is usually used for the dough and will give a product with a softer bite.

15.2.2 Mixing and Fermentation

The sponge and dough are usually mixed on vertical spindle mixers. Two or three spindle machines are used with slow mixing speed, 25 rpm. The slow and gentle mixing action incorporates the ingredients well without undue work input at the sponge stage (Figs 15.6 and 15.7).

Fig. 15.6 Vertical mixer with three spindles and 1000 kg capacity from Dingson Food Machinery.

The sponge is mixed as an 'all in' mix. The yeast should be dispersed in water before feeding to the mixing bowl. The dough is mixed to a temperature of 30–35°C, which is the optimum temperature for the action of the yeast. The sponge is fermented for 18 h at a temperature of 30–35°C and an RH of approximately 80%. During this time, the pH value will change from about 5.8 to 4.0, and the temperature of the sponge will increase.

After the fermentation of the sponge, the dough tub is taken back to the mixing room. The additional ingredients for the dough are added, including the sodium bicarbonate. Gentle, slow-speed mixing is required until a homogeneous dough is made. Overmixing will reduce the spring and give

Fig. 15.7 Vertical spindle mixer from Apinox Srl. This picture belongs to Apinox Srl—Italy. This mixer has been specifically designed for the production of soda cracker, cream cracker with two stages, fermented dough and delicate dough.

a hard, tough product. After mixing, the dough is returned to the fermentation room for up to 6 h. With the addition of the soda, a large change in the pH occurs, and the dough will reach a pH of over 7.0.

15.2.3 Dough Forming

The dough is laminated, usually with 4–6 layers at around 4.0 mm thickness (Fig. 15.8).

Fig. 15.8 Dough laminations from a cut sheet laminator.

The dough is then gauged with a maximum reduction at each gauge roll unit of 2:1. Excessive reductions of the dough thickness will prevent good lift or spring of the cracker. Typical settings for the gauge roll gaps are the following:

Gauge roll 1	12 mm
Gauge roll 2	6 mm
Gauge roll 3	3 mm

A relaxation conveyor is used to relax the tension in the dough sheet before cutting, as the soda cracker doughs are subject to considerable shrinkage after cutting and during baking.

The dough sheet is cut with a 'scrapless' cutter. Each cracker shape is perforated (not cut through), so that the dough sheet remains complete. A small amount of edge scrap is cut off, and this is diverted by side scrap wheels to the scrap return system. An additional cross scrap conveyor is used to convey the scrap to the side return conveyor. The dough sheet may be cut through across its width with one revolution of a large diameter cutting roll so that the dough sheet is divided into lengths of approximately 1.0 m in the oven. This allows shrinkage to occur during baking without random breaks in the dough sheet, which would cause problems at the cracker breaker.

15.2.4 Baking

The baking of soda crackers normally follows the US practice with a direct gas-fired oven and heavy mesh oven band (20.5 kg/m^2). Typically, a Compound Balanced Weave CB5 band is used with preheat to give a high band temperature, over 150°C at the feed end of the oven. Heat is immediately conducted into the bottom of the dough sheet, initiating a fast and strong lift or spring to give the open, flaky texture of the cracker (Fig. 15.9).

Fig. 15.9 Direct gas-fired oven with preheat from Baker Perkins.

A considerable amount of water must be evaporated from the soda cracker dough, and this is achieved by a high temperature in the middle of the oven (minimum 300°C on a direct gas-fired oven). The fast baking time and high water evaporation require a powerful oven. Soda cracker ovens have a burner capacity of over 30 kW/m^2 of oven band area. The first zone will have a burner capacity of 45–50 kW/m^2 of oven band area.

Temperature profile	300/300/280/250°C
Baking time	2.5–3.0 min

Alternative Oven Band

Good-quality soda crackers can also be baked on a Z47 wire-mesh band with pre-heat. The process details are similar, but the baking time is longer, usually 4.0 min.

15.2.5 Oil Spray

Soda crackers are oil sprayed immediately after baking. The cracker strips are transferred from the oven stripping conveyor to the oil spray machine, where a mist of fine oil is sprayed on the top and bottom of the crackers. Coconut oil or palm kernel oil is used (Fig. 15.10).

Fig. 15.10 Oil spray machine in China.

15.2.6 Cracker Breaking

The baked sheet is broken into lateral strips by a weighted roller positioned over the oven stripping conveyor. The wire-mesh conveyor is supported below by rollers before and after the cracker breaker roll. The sheet of the crackers is depressed by the breaker roll and breaks into separate strips (Fig. 15.11).

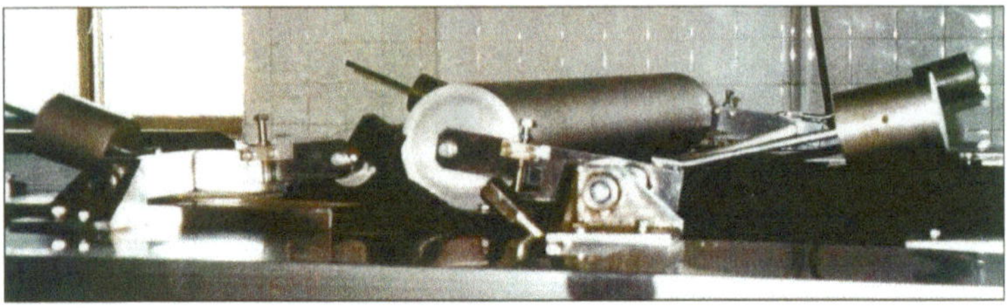

Fig. 15.11 Cracker-breaking roll at the oven end.

After cooling, the cracker strips are flip stacked by differential belt speeds between the final cooling conveyor and the packing table. After stacking, the strips must be correctly aligned. Usually side guide rolls or belts are used to nudge the strips into a central position for breaking. The strips are depressed by breaker wheels aligned with the perforations in the cracker sheet and break at the perforations. The products are normally broken into pairs.

15.3 PRODUCTION OF SEMI-SWEET BISCUITS

See Fig. 15.12.

Fig. 15.12 Marie biscuit.

15.3.1 Description

Marie is a classic biscuit made throughout Europe and Asia. It has a light, crisp, delicate texture, with pale colour and clear smooth surface.

Product Specification

Dimensions	66.0mm diameter
Thickness	6.0mm
Weight	8.3g
Appearance	Smooth surface and clear printing
Colour	Pale golden
Texture	Crisp and light
Moisture	1.5%

Marie biscuits are made with medium protein flour and SMS to develop a soft extensible dough. The doughs are mixed on horizontal mixers to a temperature of 40–42°C.

The dough is sheeted and cut and is traditionally baked on a steel band, although other semi-sweet biscuits are baked on a wire-mesh Z47 type band.

Formulation

Flour	100.00
Cornflour	4.41
Maize flour	14.70
Granulated sugar	25.59
Invert syrup 80%	7.94
Shortening	11.03
Lecithin	0.57
Salt	0.88
Soda	0.67
ACP	0.08
Protease	0.02
SMS 10% solution	0.02
Ammonium bicarbonate	0.73
Water	26.47

Critical Ingredients

1. Flour should not exceed 9.0% protein. Higher protein will result in a hard biscuit.

2. Cornflour and maize flour are used to reduce the total gluten content and make a more tender eating biscuit.
3. SMS will modify the protein to make a soft extensible dough.

15.3.2 Mixing

An 'all in one mix' on a horizontal mixer. Mixing is critical to developing the soft extensible dough. A mixing action that kneads the dough without too much tearing and extruding is ideal. Mixing time on a typical high-speed mixer will be 20–25 min. Marie doughs are mixed until the required temperature is achieved. The dough should reach 40–42°C. At this temperature, it should be well kneaded and of correct consistency for machining. Higher dough temperatures result in unstable doughs. The dough is used straight away without standing, and it is important to maintain the temperature.

15.3.3 Forming

The dough may be laminated, but doughs made with SMS are usually sheeted without lamination. Dough scrap incorporation is very important and should be very even and consistent. The temperature of the scrap dough should be as close as possible to the temperature of the new dough. Dough sheet reduction should be gentle and should not exceed the ratio of 2:1.

Typical roll gaps are the following:

Forcing roll gap on sheeter	18.0 mm
Gauging gap on sheeter	9.0 mm
First gauge roll	5.7 mm
Second gauge roll	2.5 mm
Final gauge roll	1.1 mm (cutting thickness, 1.3 mm)

The doughs shrink and require good relaxation before cutting. Separate cutting and printing rolls on the rotary cutter are recommended to achieve good, clear printing and dockering (piercing of holes in the dough pieces) (Fig. 15.13).

15.3.4 Baking

Steam may be used at the oven entry to achieve a high humidity. This will improve the surface finish of the biscuit. Baking with radiant heat transfer will give the best volume and texture for the biscuits (Fig. 15.14):

Fig. 15.13 Baker Perkins rotary cutting machine with scrap dough lifting and return conveyor.

Fig. 15.14 Steam application at oven entry.

Baking time	5.0–6.5 min
Temperatures	200/220/180°C
Moisture	Less than 1.5%

15.3.5 Cooling

A ratio of cooling to baking time should be at least 1.5:1. This will help to avoid checking.

(cracking of the biscuits after packaging due to an internal moisture gradient).

15.4 PRODUCTION OF MOULDED SHORT DOUGH BISCUITS

See Fig. 15.15.

Fig. 15.15 Glucose biscuit.

15.4.1 Description

A rotary moulded biscuit derived from 'glucose', a highly popular biscuit in India. One maker in India, Parle Products, sells approximately 13 billion of their Parle-G glucose biscuits every month. Glucose-type biscuits are made in very large volumes in India and were successfully introduced into Malaysia and Indonesia.

The biscuit is a short, sweet, rotary moulded product, developed from the English 'malted milk' biscuit. It is cheap, nutritious and satisfying. It is often fortified with vitamins and minerals, particularly calcium and iron.

Product Specification

Dimensions	58 × 37 mm
Thickness	6.7 mm
Weight	5.2 g
Appearance	Bold design
Colour	Golden
Texture	Short and tender
Flavour	Sweet
Moisture	1.2%–1.4%

Formulation

Flour (weak)	100.00
Powdered sugar	29.50
Palm oil	19.50
Lecithin	0.62
Glucose 42DE	2.50
Fructose	1.25
Skimmed milk powder	2.35
Calcium carbonate	0.55
Salt	1.10
SAP	0.30
Ammonium bicarbonate	0.58
Soda	0.49
Vitamin mix	0.11
Flavours	0.21
Water	13.33

Critical Ingredients

1. Flour should not exceed 8.0%–9.0% protein. Higher protein will result in a tough biscuit, particularly if it is overmixed.
2. The vitamin mix should be developed to suit the local requirement.

15.4.2 Mixing

The mixing process must not develop the gluten in the flour as this will result in a tough biscuit. Mixing is therefore done in two stages.

In the first stage all the ingredients are mixed except the flour and soda. The mixing continues until a consistent, homogenous cream is produced. It is important to keep the temperature of the dough low, and chilled water is required for the mixer jacket.

For the second stage, the flour and soda are added and mixed for 1 min at slow speed and 1–2 min at high speed. The final dough temperature should be 18–22°C.

Mixing may be carried out on a horizontal or a vertical spindle mixer (Fig. 15.16).

Standing Time

The dough will be sticky when discharged from the mixer as the flour has not fully absorbed the water. The dough should be stood in a cool area for 30 min before forming. It will then be less sticky and will release from the rotary moulding roll more easily.

Fig. 15.16 Peerless Sigma Arm Mixer for short and cookie doughs.

15.4.3 Rotary Moulding

The dough should be fed very evenly and consistently to the hopper of the rotary moulder, maintaining an even level across the width of the machine during the production. Adjustment will be made to the forcing roll gap, knife position and pressure roll to achieve a good release and good product shape (Fig. 15.17).

Fig. 15.17 Rotary moulding.

15.4.4 Baking

Steam may be used at the oven entry to achieve a high humidity. This will allow the biscuit to expand in the first zone and achieve good volume:

Baking time	5.0–5.5 min
Temperatures	180/200/220/200/180°C
Moisture	Less than 3.0%

15.4.5 Cooling

Sufficient cooling is required to set the biscuit, which will be soft as it leaves the oven, usually 1:1.5 baking to cooling time (Fig. 15.18).

Fig. 15.18 Two-tier cooling system from Baker Perkins.

15.5 PRODUCTION OF A CHOCOLATE CHIP COOKIE

See Fig. 15.19.

Fig. 15.19 Chocolate chip cookie.

15.5.1 Description

Short cookies with inclusions of chocolate chips or nuts.

Product Specification

Dimensions	55 mm diameter
Thickness	12.0 mm
Weight	15.0 g
Appearance	Round, irregular shape with chips visible
Colour	Golden brown
Texture	Short
Flavour	Rich with chocolate or nut flavour
Moisture	2.5–3.0%

Formulation

Flour	100.00
Shortening	55.98
Granulated sugar	50.05
Brown sugar	0.76
Whole egg powder	1.24
Vanillin	0.10
Invert syrup	1.24
Salt	0.96
Ammonium bicarbonate	0.29
Sodium bicarbonate	0.67
Chocolate chips	30.00
Water	19.14

15.5.2 Mixing

The mixing is in two stages on a horizontal or a vertical mixer. Good dispersion of the fat over the flour particles is important, and there must be a good proportion of solid fat.

The following ingredients are mixed gently at the first stage: shortening, sugars, water, salt, egg powder, vanilla, invert syrup and ammonium bicarbonate. These ingredients are mixed to dissolve the sugar and achieve a creamy emulsion. The water should be cold and the mix kept as cool as possible.

The flour and sodium bicarbonate are added for the second stage. The mixing is continued at low speed for no more than 1 min to obtain a homogenous mixture without hydration of the flour and formation of the gluten. The chocolate chips or nuts are added close to the end of the mix and given enough time to disperse evenly through the dough.

15.5.3 Forming

The dough is fed to the hopper of a wire-cut machine. The dough may be fed from a bowl by gravity. The feed rolls of the depositor operate continuously and will extrude the dough through the dies. As the dough is extruded, it is cut by a horizontally reciprocating wire. The cut dough pieces drop directly on to the baking tray or oven baking band (Figs 15.20–15.22).

Fig. 15.20 Baker Perkins wire-cut machine.

Fig. 15.21 Filler block withdrawal.

Fig. 15.22 Dough pieces being cut and deposited onto the oven band.

15.5.4 Baking

Baking on an indirect radiant (cyclotherm) oven is ideal to give good spread on the oven band, good texture and flavour. 'Recirc' ovens are widely used in North America:

Baking time	7.0 min
Temperatures	180–220°C
Moisture	2.5–3.0%

15.5.5 Cooling

A ratio of cooling to baking time should be 1:1.5.

BIBLIOGRAPHY

Almond, N., 1989. Biscuits, Cookies and Crackers. vol. 2 Elsevier Applied Science.

Davidson, I., 2016. Biscuit Baking Technology, second ed. Academic Press, Elsevier.

Manley, D., 1996. Technology of Biscuits, Crackers and Cookies, second ed. Woodhead Publishing Ltd.

Sykes, G., 2016. Baker Pacific Ltd.

CHAPTER 16

Ingredients for Biscuits: An Introduction

The main ingredients for biscuit making are flours, sugars and fats. To these ingredients, various small ingredients may be added for leavening, flavour and texture.

16.1 FLOUR

16.1.1 Wheat flour

The principle ingredient of biscuits is wheat flour. The grain consists of bran (12%), which is the outer husk; endosperm, the white centre (85.5%); and tiny germ (2.5%). Typical biscuit flour is milled to a yield or extraction of 70%–75%. Whole-meal flour is of 100% extraction, and wheat meal flours in between these extraction rates are normally around 84% extraction. The flour will also contain moisture of between 13% and 15%.

The wheat flour is composed of carbohydrate (as starch), protein and fat, together with some fibre, ash and trace minerals and vitamins. The protein is mainly gluten, composed of gliadin and glutenin.

The percentage of protein determines the flour strength. A dough made from strong flour with a high protein content is extensible and can be machined into a continuous sheet for crackers and hard biscuits. A weak flour with a low-protein content produces a short dough that may be moulded or a soft, high-fat dough that may be deposited on the baking band and when baked gives a soft, tender cookie (Table 16.1).

16.1.2 Wheat Gluten

The formation of the gluten, its strength and elasticity are largely determined by the flour specification, recipe and the mixing and forming processes. Wheat flour contains proteins including gliadin and glutenin. In the presence of water, these proteins combine to form gluten. As the dough is mixed, the protein molecules form long strands of gluten, which have strength and elasticity. The gluten forms an elastic web, which gives the

Biscuit, Cookie and Cracker Production
https://doi.org/10.1016/B978-0-12-815579-0.00016-7

Table 16.1 Typical composition of weak/medium and strong flours

	Weak/medium flour	Strong flour
Starch	74.5	71.5
Moisture	14.0	13.5
Proteins (gluten forming)	7.0	10.0
Proteins (soluble)	1.0	1.0
Sugar	2.0	2.5
Fat	1.0	1.0
Total	100.0	100.0

dough strength and allows it to be machined into a thin sheet for crackers and semi-sweet biscuits. These biscuits are made with relatively 'strong' flour, which has a high-protein content, typically 10%–12%.

The gluten web is also important in trapping air and gas bubbles formed by yeast fermentation or by leavening agents such as sodium bicarbonate ('soda') or ammonium carbonate ('vol'). This leavening process, combined with the laminating of the dough, gives the characteristic open, flaky texture of crackers during baking.

Soft or short biscuits are generally made with low-protein flour (7%–9%). A low-protein flour makes a dough with a much weaker gluten web. In addition, these doughs have higher fat contents. The fat coats the flour particles, and this inhibits the hydration of the proteins and the formation of the gluten web. Shorter mixing times also result in less development of the gluten strands, and hence, the biscuits have a short texture.

16.1.3 Starch

Starch is the main component of wheat flour. It represents almost all of the carbohydrate content and around 80% of the total energy content of wheat flour. Starch is a polysaccharide (many sugars) made up of glucose units linked together to form long chains. The principle starch molecules in wheat flour are amylose, which typically comprises 28% of the total amount of starch. Amylose molecules contribute to gel formation. Their linear chains of molecules line up together and are able to bond to make a viscous gel.

Starch is insoluble in water. However, the starch granules do absorb a limited amount of water in the dough and swell. Above temperatures of 60–70°C, the swelling is irreversible, and gelatinisation begins.

The gelatinisation may continue until the starch granules are fully swollen, but it is normal in baked products that only partial gelatinisation occurs. The gelatinisation of the starch contributes to the rigidity and texture of the biscuit.

As the starch gel is heated further, dextrinisation occurs. This contributes to the colouring of the biscuit.

In soft dough products, the high sugar and fat content of the dough inhibits starch gelatinisation. The presence of sugars delays the gelatinisation of the starch, which may be due to the competition for water. The fat, composed of triglycerides and surfactants, also tends to inhibit gelatinisation. With high sugar and fat recipes, the dough has a low gel viscosity and strength and produces short and soft biscuits and cookies.

16.1.4 Corn Flour

Corn flour is a white free-flowing powder produced by wet milling of maize, followed by washing, concentrating, centrifuging, drying, milling and sifting to give a natural maize starch. It has a short gel texture and relatively high viscosity and is easily dispersed in cold water.

The protein in corn flour does not form gluten, and it can be used as an ingredient to produce a more tender biscuit with reduced gluten development.

16.1.5 Oats

Oatmeal and oat flakes or rolled oats are used in various biscuits providing a distinctive flavour and texture. Oatmeal flour is a coarse flour used with wheat flour to produce short, friable biscuits. Oat flakes are a good addition for a range of cookies.

16.1.6 Soya

Soya flour is made from soya beans after the oil has been extracted. Soya flour is a source of protein and contains a small amount of the emulsifier lecithin. Soya flour is a valuable ingredient in some dietary and gluten-free biscuits.

16.2 SUGARS

16.2.1 Sucrose

Common sugar (sucrose) is a carbohydrate derived from sugar cane or sugar beet. It is a disaccharide composed of two monosaccharides, a molecule of glucose joined to a molecule of fructose. Sugar is used in biscuit formulations in a granulated or powder form.

Sugar gives sweetness, but it is also important in developing the texture of the biscuit. Dissolved sugar tends to inhibit starch gelatinisation and gluten formation and creates a biscuit with a more tender texture. Undissolved sugar crystals give a crunchy, crisp texture. Sugar crystals, which melt during baking, cool to a noncrystalline glasslike state that gives a crispy, crunchy texture, particularly on sugar-topped biscuits.

Dry sucrose melts at 160–186°C. Biscuits with sugar toppings that are melted to a smooth, shiny surface require high-intensity flash heat at the end of the oven to fully melt the sugar.

Invert sugar syrup is a mixture of glucose and fructose. The sucrose is split into its component monosaccharides by hydrolysis. The sucrose in solution is heated with a small quantity of acid such as citric acid. After inversion, the solution is neutralised by the addition of soda. The invert syrup is sweeter than sugar, and it contributes to a moist, tender texture in the biscuit.

Other reducing sugars are included in biscuit formulations in syrup form, for example, glucose syrup, malt extracts and honey. The reducing sugars in the presence of amino acids produce the Maillard reaction that contributes to the colour of the biscuit.

Sugar specifications:

Moisture content	0.06% maximum
Ash content	0.03% maximum

Particle size (average):

Powdered sugar	60 μm
Crystal sugar	150 μm
Caster sugar	150–450 μm
Granulated sugar	250–1050 μm

Brown sugar, a dry golden brown sugar with bold crystals.

Particle size	0.8–1.2 mm

16.2.2 Glucose Syrup

Glucose syrup ($C_6H_{12}O_6$) is a solution (up to 80%) of glucose (dextrose), maltose and maltodextrins in water. It is normally obtained by enzymatic hydrolysis of starch. Starch from wheat, corn, potato, cassava or any other plant can be used for this purpose.

Standard glucose syrup has a dextrose equivalent (DE) value of 42. The relative sweetness of 42 DE glucose to sucrose is 40%–45%.

16.2.3 Cane Syrup 80%

Syrups with 80% solids derived from the refining of cane sugar are used for their excellent flavour.

16.2.4 Invert Syrup 70%

Syrup of 70% solids made by acid hydrolysis of sucrose. The result is a 50:50 mixture of dextrose and fructose that are both reducing sugars and contributing to the Maillard reaction during baking.

16.2.5 Fructose Syrup 80%

Commercially, fructose is usually derived from sugar cane, sugar beets and corn. Crystalline fructose is a monosaccharide, dried and ground and of high purity. High-fructose corn syrup (HFCS) is a mixture of glucose and fructose. High-fructose syrup is a special type of glucose, formulated with a high-fructose content. It is much sweeter than glucose.

16.2.6 Malt Extract 80%

A thick glutinous syrup of 80% solids usually nondiastatic and obtained by water extraction of malted wheat or barley. The heat treatment used to concentrate the solution destroys any enzymes. It is used as an important flavour ingredient. It is rich in maltose, which is a reducing sugar.

16.3 DOUGH FATS AND OILS

16.3.1 Vegetable Fats

Fats are a vitally important ingredient in achieving the texture, mouth feel and bite of the biscuit. The fats are produced from good-quality crude oils by a process of refining, bleaching and deodorising. They are produced primarily from vegetable oils but may contain hydrogenated fish oils.

Typical blended vegetable dough fats are solid at ambient temperature and melt over a wide temperature range. Most fats used in biscuit making are melted below blood temperature (36.9°C), and this avoids a waxy mouth feel. Fats are specified with a solid fat index (SFI), which indicates the percentage of solid fat present in the total fat. A vegetable shortening typically has an SFI of around 21% at 25°C and 17% at 30°C (Fig. 16.1).

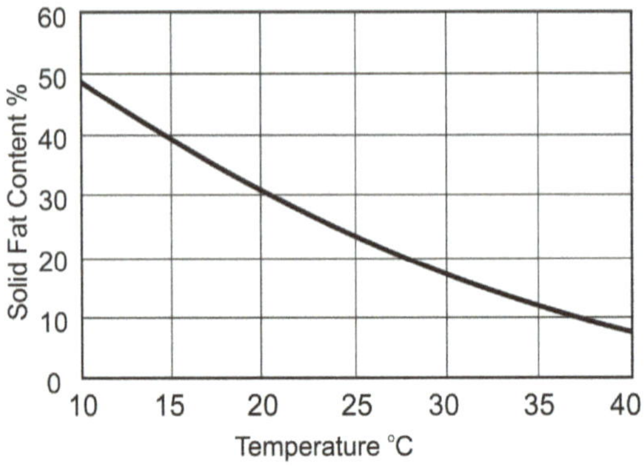

Fig. 16.1 Graph showing melting profile (solid fat index vs. temperature) of a typical vegetable fat for biscuit dough.

Recipes with high-fat contents require little water for producing a cohesive dough and produce soft, short doughs. During mixing, the fat coats the flour particles, and this inhibits hydration and interrupts the formation of the gluten. Fats also tend to inhibit the leavening action of the carbon dioxide diffusion in the dough during baking, and this produces a softer, finer texture. Where both fat and sugar amounts in the recipe are high, they combine to make a soft, syrupy, chewy texture.

Typically, crackers and hard sweet biscuit doughs, which are sheeted and cut, have fat contents of 10%–22% of the flour by weight. Rotary moulded doughs may have 17%–30% of fat and wire-cut and deposited cookie doughs 25%–60%.

16.3.2 Butter

Butter is used for its shortening and flavour. The flavour of the butter is complemented by sugar and vanilla during baking and gives a distinctive flavour and aroma.

16.3.3 Coconut Oil

Coconut oil is typically used for oil spraying. The oil is hydrogenated, neutralised, deodorised and bleached. Melting point, 32–34°C.

16.4 OTHER INGREDIENTS

16.4.1 Whole Egg Powder

Whole egg powder is spray dried. Egg yolk is rich in fat and lecithin, and it is these ingredients that enhance the flavour and eating quality of the cookie.

16.4.2 Lecithin

Lecithin is an emulsifier produced from soya beans and available in liquid or powder form. It may be added to the fat or directly into the dough mix.

16.4.3 Yeast (Fresh)

Yeast is a microscopic, unicellular organism. It breaks down sucrose and maltose into monosaccharides and glucose and fructose into alcohol and carbon dioxide. During fermentation, the gas production causes the dough to increase in volume and develop the characteristic flavour. Fresh yeast is available as a compressed block. Active dried yeast and instant active dried yeast are other forms.

16.4.4 Ammonium Bicarbonate ('Vol') $(NH_4)HCO_3$

A volatile salt is an effective leavening agent. When heated, it liberates carbon dioxide, ammonia gas and water.

16.4.5 Sodium Bicarbonate ('soda') $NaHCO_3$

Sodium bicarbonate is the most important aerating agent. When heated, it reacts with acidic materials in the dough to release carbon dioxide and water.

16.4.6 ACP—Acid Calcium Phosphate

Acid calcium phosphate is also known as monocalcium phosphate. It is used as a leavening agent in conjunction with sodium bicarbonate and ammonium carbonate.

16.4.7 SAPP—Sodium Acid Pyrophosphate

Sodium acid pyrophosphate is commonly used in the baking industry as a leavening agent. It combines with sodium bicarbonate to release carbon dioxide.

16.4.8 Salt (NaCl)

Sodium chloride is used as a flavour enhancer and also to control the rate and extent of fermentation.

16.4.9 SMS—Sodium Metabisulphite $Na_2S_2O_5$

Sodium metabisulphite is a reducing agent for the modification of the strength of the gluten in doughs. It causes the gluten to become more extensible and less elastic and so reduces shrinkage of the dough pieces during baking.

16.4.10 Proteolytic Enzyme

Proteinases break down and modify the gluten in doughs, giving a softer, less elastic dough.

16.4.11 MRP—Material Recovered From Production

Many biscuits contain rework such as damaged biscuits recovered from production. Hygiene in collection, storage and processing is essential. The biscuits are ground and may then be added to some recipes at up to 10% of the total.

BIBLIOGRAPHY

Baking Management, 2008. A Tropical Touch. Soy-based Solutions. Penton Media Inc. http://www.baking-anagement.com/ingredients/brief_guide_functional_1028/index1.html.

Benedict M. n.d. University of Houston. How Does Temperature Affect Yeast Activity? MadSci Network. http://www.madsci.org/posts/archives/jan2001/980908832.Gb.r.html.

British Sugar plc, 2015. Screened White Sugars. www.britishsugar.co.uk.

Buck, J.S., Walker, C.E., 2009. Sugar and Sucrose Ester Effects on Maize and Wheat Starch Gelatinisation Patterns. Starch. Wiley VCH Verlag GmbH.

Dakota Yeast, 2010. Yeast Fermentation in Baked Goods. http://www.dakotayeast.com/help-fermentation.html.

DTKFCPL, 2003. Examining the Role of Fats in Bakery. April 03.

Eyre, C., 2008. AB Enzymes Launch Targets Improved Biscuit Baking. Decision News Media. www.bakeryandsnacks.com.

Flour Specifications, 2015. www.flourmilling.co.uk.

Fineli®, 2003–2010. National Institute for Health and Welfare. Wheat Flour Whole Grain.

Food Resource, 2010. Oregon State University. Bread Dough. Starch.

Food Resource, 1990. Oregon State University. Gel. Excerpts from Bender Arnold E. Dictionary of Nutrition and Food Technology, Butterworths, Boston.

Ghiasi, K., Hoseney, R.C., Varriano-Marston, E., 1981. Effects of flour components and dough ingredients on starch gelatinisation. Cereal Chem. 60 (1), 58–61.

Gurney, A., 2008. Bakery Fats and Oils. Leading Edge.

Lansbergen, G., 2002. Fats for Food Consultants. http://www.fatsforfoods.com/specifications.htm.

Lowe, B., 2009. Experimental Cookery from the Chemical and Physical Standpoint. Gluten.

Manley, D., 2001. Biscuit, Cracker and Cookie Recipes for the Food Industry. Woodhead Publishing Ltd.

Manley, D., 1996. Technology of Biscuits, Crackers and Cookies, second ed. Woodhead Publishing Ltd.

Manley, D., 1998. Manual 1, Ingredients. Wood Head Publishing Ltd.

Moodie, P., 2001. In: Traditional baking enzymes—proteases. Enzyme Development Corporation, Presented at American Institute of Baking, Manhattan, Kansas.

Toufeili, I., Ismail, B., Shadarevian, S., Baalbaki, R., Khatar, B.S., Bell, A.E., Schofield, J.D., 1999. The role of gluten proteins in the baking of Arabic bread. J. Cereal Sci. 30, 255–265.

Wade, P., 1998. Biscuits, Cookies and Crackers. vol. 1. Elsevier Applied Science Publishers Ltd.

CHAPTER 17
Quality Control: An Introduction

17.1 SCOPE OF QUALITY CONTROL

Quality control covers checks of raw materials, manufacturing process and finished products and packaging. Quality control involves all areas of the company activities: purchasing, warehousing, production, engineering, sales and laboratory testing.

There are several standards applied in the biscuit industry and recognised internationally: HACCP, ISO 9000 series, BS 570.

Quality control functions in a biscuit bakery will cover the following elements:

- Specifications and quality of all raw materials including ingredients and packaging materials
- Production and packaging equipment
- Equipment maintenance
- Process control throughout the manufacturing process
- Production output, downtime and damaged product
- Biscuit dimensions, colour, weight, texture and flavour
- Pack weights, appearance, labelling and security
- Metal detection
- Factory conditions, hygiene, cleaning procedures and services
- Warehousing, storage and despatch of the biscuits
- People and their facilities

There is now available variety of software for quality control management in the food industry. A range of software can be found at www.capterra.com.

17.2 PROCESS AUDIT

A process audit may be made for each product detailing the machine settings, process times and temperatures. This may be used by the production staff during each shift and all variations, faults, problems and downtime recorded. This will include any adjustments to the recipe and mixing process, settings

Biscuit, Cookie and Cracker Production
https://doi.org/10.1016/B978-0-12-815579-0.00017-9

of the forming machines and baking time and temperatures. Regular process audits will highlight problems and lead to more consistent production.

17.3 EQUIPMENT FOR QUALITY CHECKS OF INGREDIENTS AND PROCESS

Equipment and software are now available for in-line checking of several process factors such as moisture content, dough sheet thickness and biscuit counts at the oven exit. However, most checks are made off line in the laboratory or test bakery.

17.3.1 Ingredient Analysis

Advanced near-infrared spectroscopy (NIRS) provides rapid analysis for raw materials, for example (Fig. 17.1):

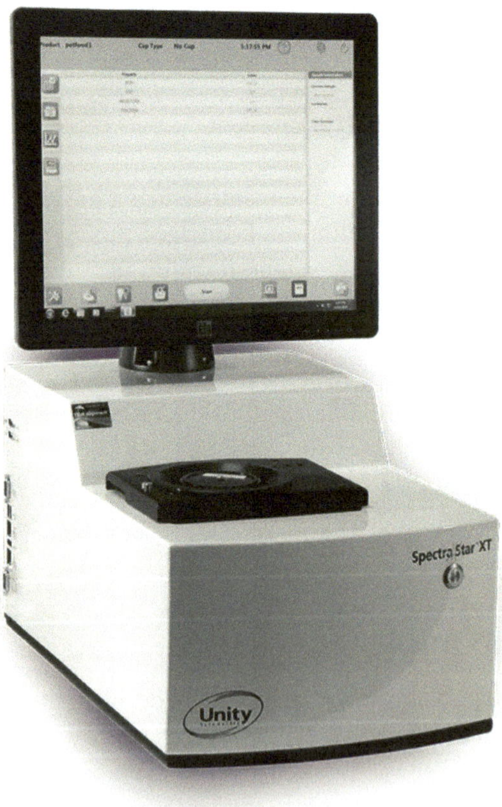

Fig. 17.1 Unity SpectraStar near-infrared spectroscope for ingredient analysis.

Flour: moisture, protein, ash, water absorption, starch damage, extensibility and falling number.
Chocolate: fat, moisture, protein and sugar.

17.3.2 Measurement of Protein

Fully automatic Kjeldahl digestion units are widely used in laboratories for nitrogen analysis and protein determination in foods (Fig. 17.2).

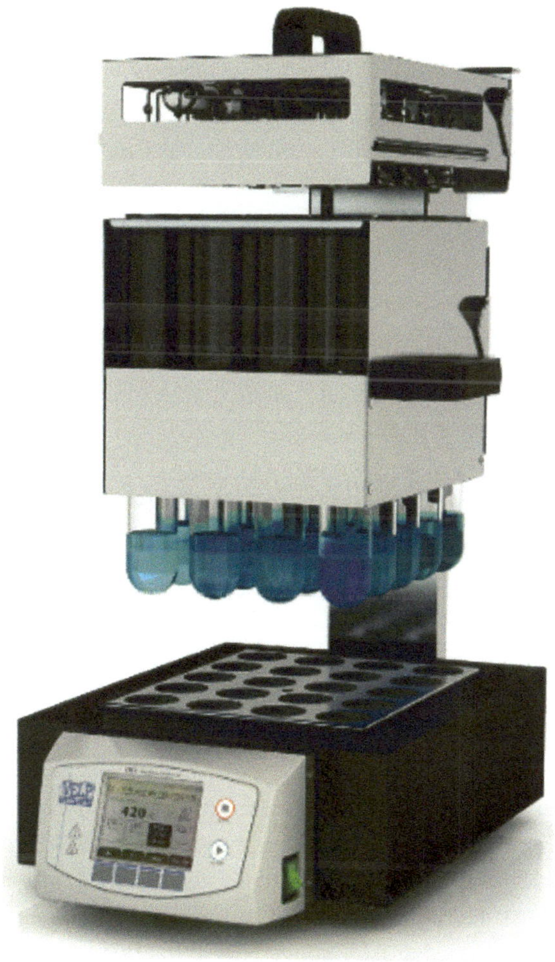

Fig. 17.2 Velp DKL20 digestion unit for analysis of nitrogen and protein.

17.3.3 Measurement of Solid Fat Index for Fats

The solid fat index (SFI) is a measure of the solid fat present compared with the liquid phase at a particular temperature. The temperature curve is related to the melting range of the fat and its properties during the process and the performance in the final product. It shows the amount of fat that has melted at a given temperature. The SFI can be determined by dilatometry. An alternative method now widely used is differential scanning calorimetry (DSC) (Fig. 17.3).

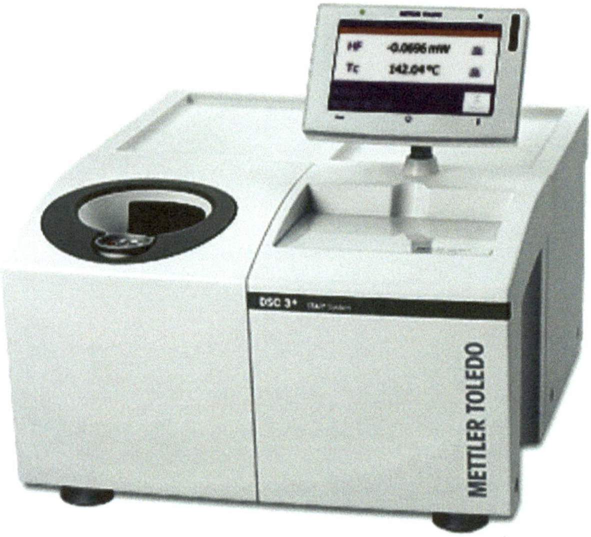

Fig. 17.3 Differential scanning calorimeter from Mettler Toledo.

17.3.4 Moisture Content

Checking the moisture content of ingredients and biscuits by near-infrared spectroscopy (Figs 17.4–17.6).

17.3.5 Refractive Index

It is important to ensure that sugar solutions used in the recipe have consistent concentrations. The refractive index (degree Brix) indicates the concentration of solutions of sucrose, glucose and invert syrup at a particular temperature. One degree Brix is 1 g of sucrose in 100 g of solution. Usually, liquid sugar is purchased at 67% solids. The refractive index is measured by a refractometer (Fig. 17.7).

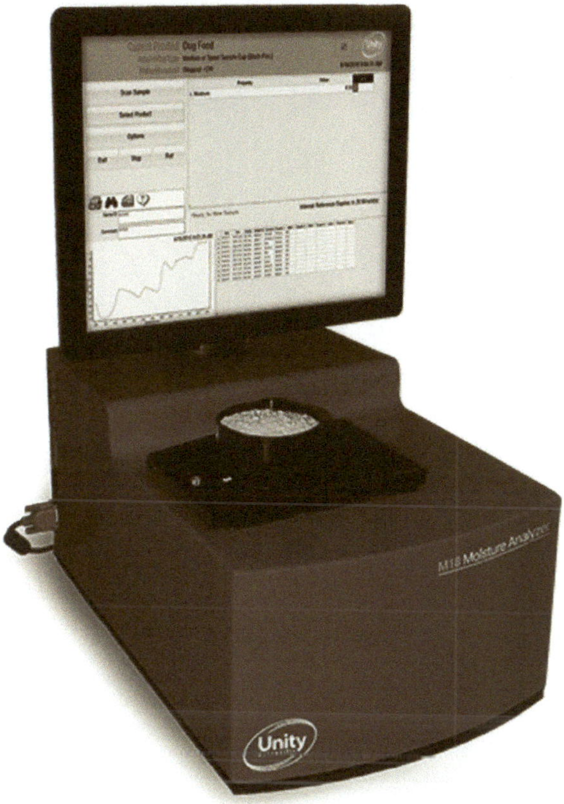

Fig. 17.4 Unity M18 moisture analyser from Unity Scientific.

Fig. 17.5 Digital moisture balance from CSC Scientific.

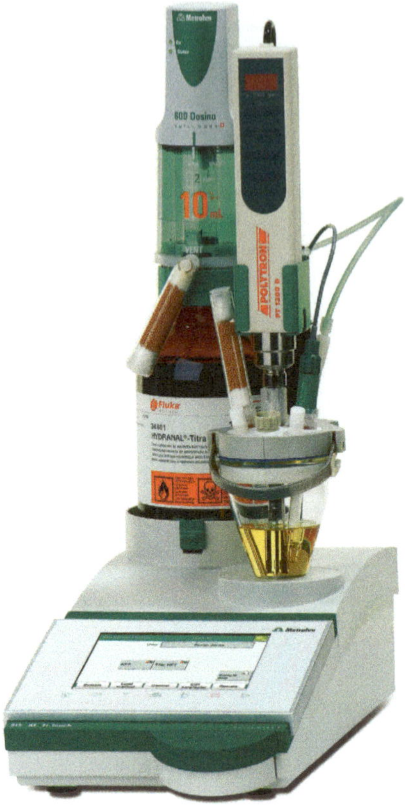

Fig. 17.6 Compact, stand-alone volumetric or coulometric Karl Fischer titrator with integrated touch-screen control unit for routine applications from Metrohm AG. www.metrohm.com.

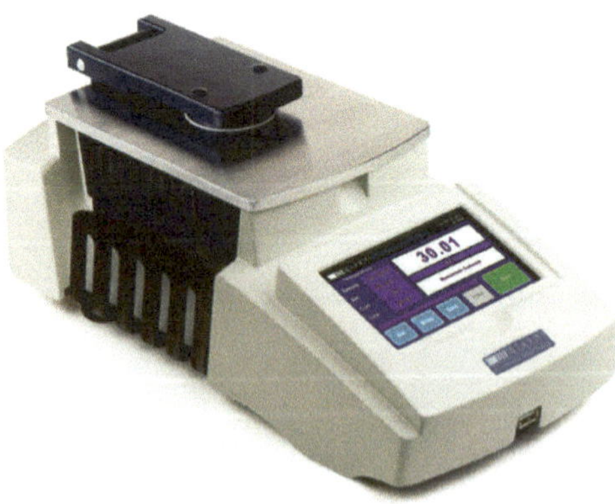

Fig. 17.7 Laboratory refractometer from Rudolph Research Analytical.

17.3.6 Colorimeter

See Fig. 17.8.

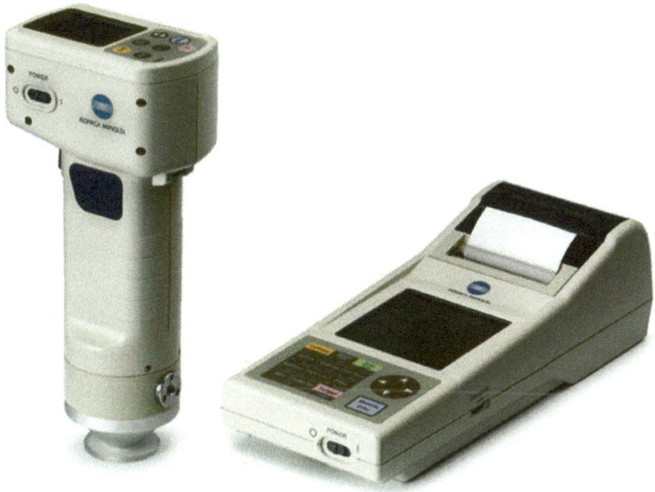

Fig. 17.8 Konica Minolta CR-410/400 tristimulus colorimeter.

17.3.7 pH Meters

pH metres measure the acidity/alkalinity of liquids and are used to test doughs. Most doughs and biscuits have a pH value of around 7 (Fig. 17.9).

17.3.8 Laboratory Balances

Analytical balances for weighments of small ingredient quantities and biscuit samples (Fig. 17.10).

17.3.9 Thermometers

See Figs. 17.11–17.13.

17.4 BAKED BISCUIT CHECKS

Samples of baked biscuits may typically be checked at the exit of the oven and recorded every 30 min:

- Biscuit weight
- Dimensions: length, width and diameter
- Colour top and bottom
- Moisture content
- Monitoring sprayed oils and cheeses

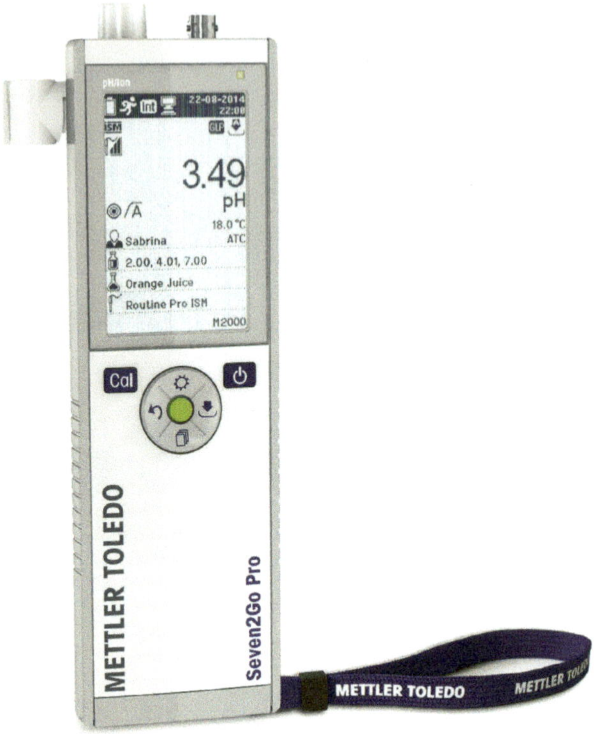

Fig. 17.9 Hand-held pH metre from Mettler Toledo.

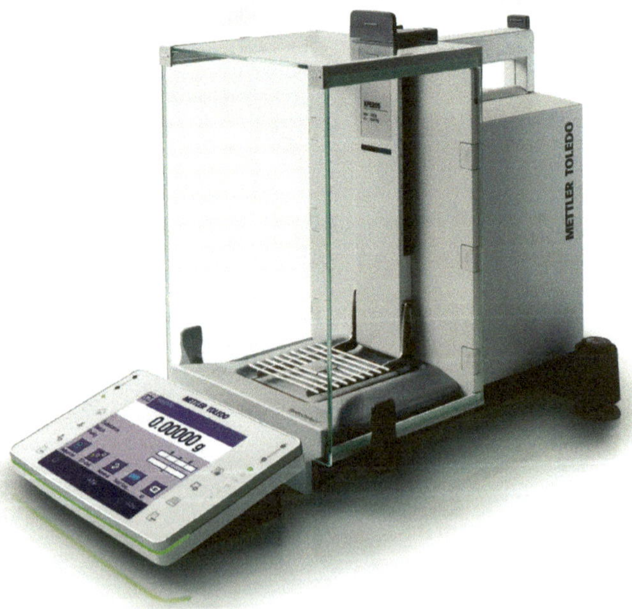

Fig. 17.10 Laboratory balance from Mettler Toledo.

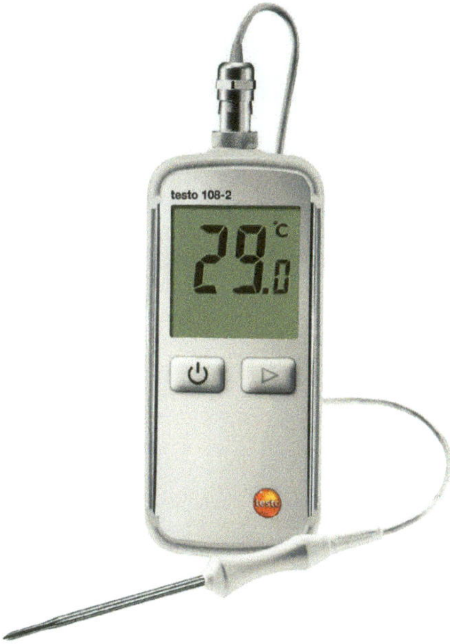

Fig. 17.11 Testo 108 digital food thermometer (waterproof).

Fig. 17.12 Testo 176T four-channel data logger.

17.5 PACKAGING CHECKS

- Pack weight variation, minimum/maximum
- Pack security
- Pack appearance and labelling
- Metal detection
- Packaging seal integrity

Fig. 17.13 ICL Calibration Laboratories thermometer.

17.5.1 Metal Detection

See Fig. 17.14.

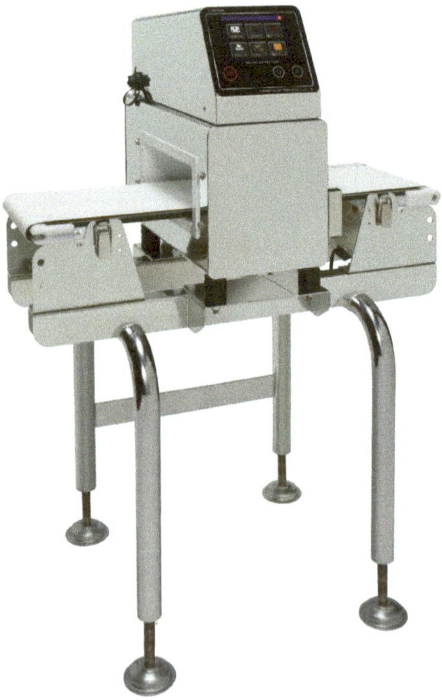

Fig. 17.14 Cosmo-one 0528 metal detector from Ishida Co. Ltd. Capable of detecting any metal from stainless steel to copper.

17.5.2 X-Ray Inspection

Integrated X-ray systems detect contamination, product defects, biscuit count and nonmetal objects such as glass and plastic (Figs 17.15 and 17.16).

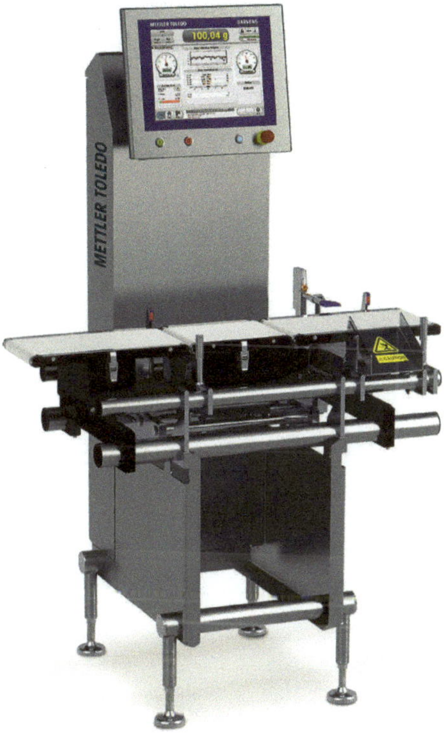

Fig. 17.15 Mettler Toledo C35 Advanced Line. The Mettler Toledo C35 is a line with integrated inspection technologies. Available as a combination system with a choice of vision, x-ray and metal detection technologies.

17.5.3 Systems for Testing Package Integrity

VeriPac inspection systems are deterministic test methods for package integrity testing that produce reliable and robust quantitative test data.VeriPac systems can be easily integrated into the packaging process to improve quality, reduce waste and provide operators with a clearer understanding of package quality (Figs 17.17 and 17.18).

FlexPak tests seals and package integrity for pouches, bags, trays and gas-flushed packs for all types of food packaging.

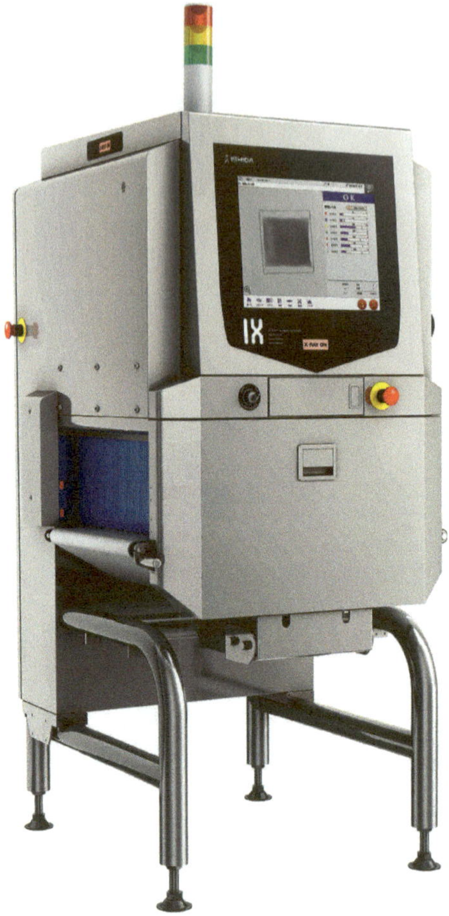

Fig. 17.16 IX-G Series X-ray inspection system from Ishida Co. Ltd.

Fig. 17.17 VeriPac inspection system from Packaging Technologies and Inspection (PTI), the United States.

Fig. 17.18 FlexPak Leak Detector with vacuum attachment package—package integrity testing (bubble emission) equipment.

BIBLIOGRAPHY

Almond, N., 1989. Biscuits, Cookies and Crackers. vol. 2. Elsevier Science Publishers.

Capterra, 2017. www.capterra.com.

CCIT, 2017. www.ccit.com.

CSC Scientific Co. Inc, 2017. www.cscscientific.com.

Extech. n.d. www.extech.com.

Flex Leak Detectors Inc, 2017. www.flexpakinc.com.

ICL Calibration Laboratories Inc, 2017. www.icllabs.com.

Konica Minolta, 2017. www.konicaminolta.com.

Manley, D., 1996. Technology of Biscuits, Crackers and Cookies, second ed. Woodhead Publishing Ltd.

Metrohm, 2017. www.metrohm.com.

Mettler Toledo, 2017. Determining the Solid Fat Index. www.mt.com.

PTI inspection systems, 2017. Packaging Integrity/Seal Integrity/Leak Detection. www.ptiusa.com.

Rudolph Research Analytical, 2017. www.rudolphresearch.com.

Testo Ltd, 2017. www.testolimited.com. www.testo.com.

Unity Scientific, 2017. Quality Control in a Cookie and Cracker Plant. www.unityscientific.com.

Velp Scientifica Srl, 2017. www.velp.com.

CHAPTER 18

Test Bakery Equipment

18.1 THE REQUIREMENT

Bakeries need equipment for the small-scale production and development of all types of biscuits and snack cakes:

- Butter cookies
- Deposited and wire-cut cookies
- Rotary moulded products
- Hard sweet biscuits
- Crackers
- Sandwich biscuits
- Snack cakes

A well-equipped laboratory and test bakery will enable the baker to develop new products, to successfully modify formulations as ingredient costs alter and to check and develop the manufacturing processes. The particular equipment specified has been used extensively over many years.

18.2 THE EQUIPMENT

18.2.1 Mixing

Two mixers are included to produce all types of biscuit and cracker doughs, soft cookie doughs, batters for cakes and creams.

18.2.2 Dough Piece Forming

A dough sheeter is used to sheet the dough to the required thickness for cutting. Cutting and moulding are carried out using hand cutters and moulds.

18.2.3 Baking

Baking is on a deck oven. Two decks are required, one with a standard crown height and heating and the other has a low crown height and additional power for baking crackers.

Biscuit, Cookie and Cracker Production
https://doi.org/10.1016/B978-0-12-815579-0.00018-0

18.2.4 The Test Bakery Facility

The facility will ensure that the bakery can produce and develop all types of product. It is an ideal facility for developing texture, shape, appearance and flavour prior to trials on a larger baking line.

18.3 SPECIFICATION OF THE EQUIPMENT FOR THE TEST BAKERY

18.3.1 Platform Scales

For the weighing of main ingredients, flour, etc. (Fig. 18.1).

Fig. 18.1 Platform scales from Brecknell.

WS 300 light capacity platform scales. With 1 part in 6000 resolution, these platform scales give a weighing performance of 300 kg × 50 g.

18.3.2 Digital Scales for Small Ingredients

For the accurate weighment of small ingredients (Fig. 18.2).

Accuracy—within ±3 divisions at full load and readability to 1 mg
Capacity—6 kg construction—ABS plastic housing and stainless steel platter
Display—easy-to-read LED
Power—AC and rechargeable battery for portability
Operator keys—tare/units and on/off/zero

Fig. 18.2 MBS scales from Brecknell.

18.3.3 Ingredient Screener

See Fig. 18.3.

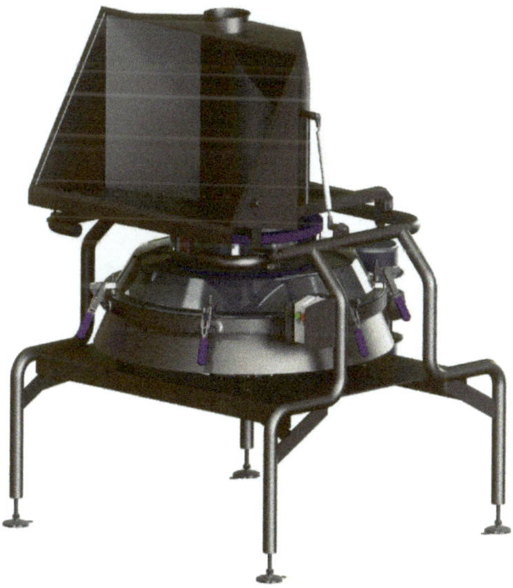

Fig. 18.3 SM Easilift screener for powder and ingredients from containers or sacks. The Easilift range is robust and powerful with full dusthood protecting operators and a unique power lift system by Farleygreene, the United Kingdom. www.farleygreene.com.

18.3.4 Hobart Mixer

Suitable for soft doughs, batters, creams, etc. (Fig. 18.4):

- A highly versatile mixer with planetary action
- Easy to operate and to clean
- Interchangeable mixing tools (beater and wire whip) for different applications
- Swing-out bowl and quick release agitators
- Three-speed gearbox and variable-frequency drive with soft start
- Overload and thermal protection
- Built-in timer with auto time recall

Fig. 18.4 Hobart countertop mixer.

18.3.5 Morton Duplex Mixer

A 'Z' blade mixer is suitable for doughs (particularly cracker, biscuit and hard sweet doughs requiring development), pastry doughs, etc. (Fig. 18.5):

Material: stainless steel for all contact parts.

Capacity: 3–5 L.

Two-speed drive

Safety: fully interlocked.

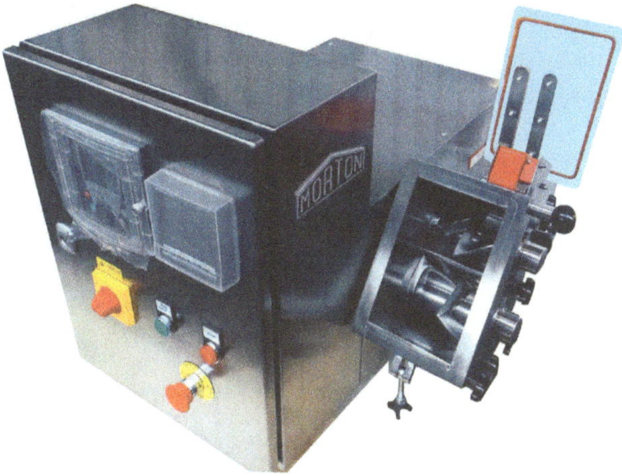

Fig. 18.5 Morton laboratory mixer.

18.3.6 Dough Sheeter

For the production of dough sheets for cracker and semi-sweet biscuit doughs. The dough sheet may be gradually reduced to the required thickness for cutting by multiple passes through the gauging rolls. The conveyors reverse so that the dough sheet can be sheeted, folded and then passed back through the rolls (Fig. 18.6).

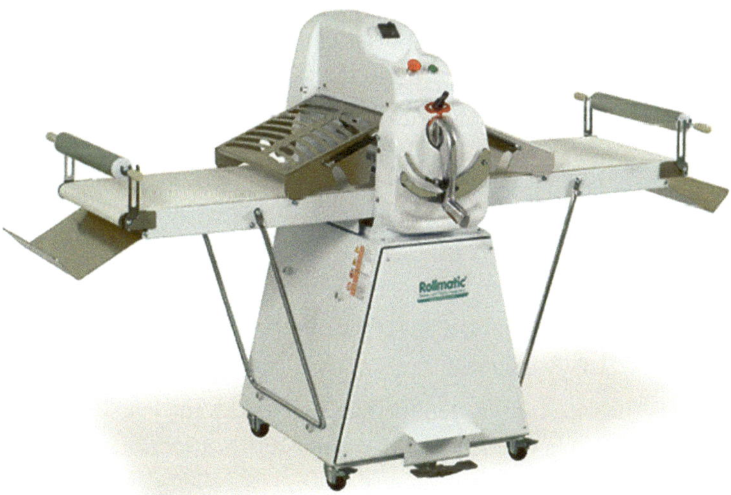

Fig. 18.6 Rollmatic Dough Sheeter SH500-600. www.rollmatic.com.

Compact sheeter on base with castors and folding tables. Control with joystick and foot (supplied as option), stainless steel cylinder guards with safety microswitch and gas spring.

18.3.7 Deck Baking Oven

Deck ovens may be heated by gas or electricity. An oven can be specified for baking all types of cookies, crackers, cakes and pastries (Fig. 18.7).

Fig. 18.7 Two-deck baking oven.

The oven should be specified with two or three decks, one deck with standard crown height and one with low crown height and high heat rating for cracker baking.

Each deck is suitable for two trays (40 × 60 cm)

Additional heat rating for crackers

Top and bottom heat regulation

High-grade insulation

18.3.8 Moisture Analyser

For checking the moisture content of biscuits, cookies and crackers (Fig. 18.8):

- Quick and reliable determination of moisture content of biscuits, crackers and cookies
- Flexibility to analyse a wide variety of samples
- Coiled quartz heating element or round ceramic IR heating element
- Easy-to-set drying programmes

Fig. 18.8 Sartorius moisture analyser MA.

18.3.9 Colour Spectrophotometer

Versatile benchtop spectrophotometer to measure the colour of solid, pasty, granular and liquid samples in either reflectance or transmittance. The top-port concept, offering various aperture sizes, allows the measurement of almost any kind of samples (Fig. 18.9).

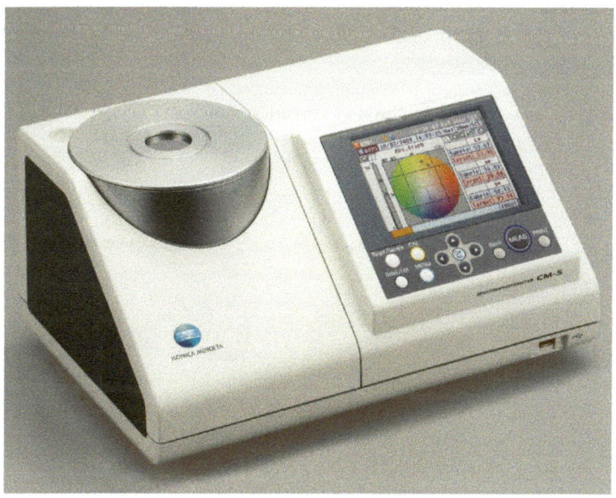

Fig. 18.9 Konica Minolta colour spectrophotometer.

18.3.10 Baking Trays

See Fig. 18.10.

Fig. 18.10 Baking trays.

18.3.11 Cookie Moulds and Stamps

See Fig. 18.11.

Fig. 18.11 Cookie stamps from Eddingtons. www.eddingtons.co.uk.

18.3.12 Cookie Cutters

See Fig. 18.12.

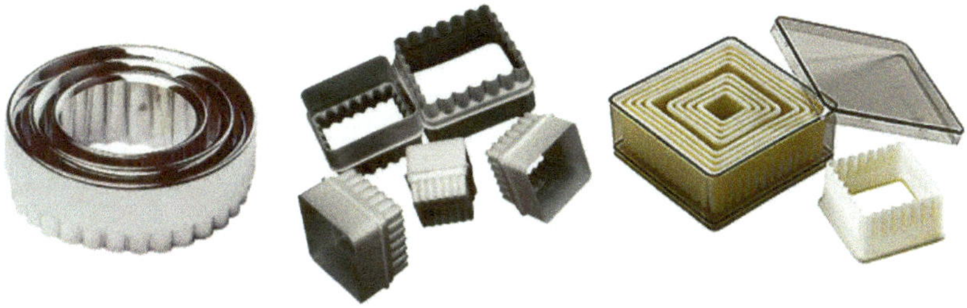

Fig. 18.12 Cookie cutters from cookiecutter.com.

18.3.13 Sieves and Sprinklers

Sieves and sprinklers with various screens for salt, sugar, seeds, cheese, etc. (Fig. 18.13).

Fig. 18.13 Sieves and sprinklers.

18.3.14 Plastic Bowls, Scrapers, Spatulas, etc.

A range of hand utensils for the bakery.

BIBLIOGRAPHY

Brecknell, 2017. www.awtxglobal.com.
Cookie Cutter, 2017. www.cookiecutter.com.
Eddingtons, 2017. www.eddingtons.co.uk.
Farleygreene, 2017. www.farleygreene.com.
Hobart Corp, 2016. www.hobartcorp.com.
Kemplex, 2017. www.kemplex.com.
Konica Minolta, 2016. www.konicaminolta.com.
Mackies, 2016. www.mackies.com.au.

Morton Mixers, 2016. www.morton-mixers.co.uk.
Nordic Ware, 2017. www.nordicware.com.
RAM srl, 2017. https://ramsrl.eu.
Rollmatic, 2017. www.rollmatic.com.
Rondo, 2017. www.rondo-online.com.
Russell Finex, 2016. www.russellfinex.com.
Sartorius AG, 2016. www.sartorius.co.uk.
Wellbake Equipment Inc, 2017. https://welbake.com.

APPENDIX 1

Suppliers of Ingredient Storage and Handling Equipment

Abbey Logistics Group
Brasenose Road, Bootle, Liverpool, L20 8HG, UK
Email: hello@abbeylogisticsgroup.com; Website: www.abbeylogisticsgroup.com
Tel. +44 (0)151 933 0999

AZO GmbH + Co. KG.
Rosenberger Str. 28, 74706 Osterburken, Industriegebiet Ost, Germany
Email: azo-solids@azo.com; Website: www.azo.com
Tel. +49 6291 92 0
Fax. +49 6291 929 500

Baker Perkins Ltd.
Manor Drive, Paston Parkway, Peterborough, PE4 7AP, UK
Email: bpltd@bakerperkins.com; Website: www.bakerperkins.com
Tel. +44 1733 283000
Fax. +44 1733 283001

Bellflow Systems Ltd.
7, Swan Business Centre, Osier Way, Buckingham MK18 1TB, UK
Email: mail@bellflowsystems.co.uk; Website: www.bellflowsystems.com
Tel. +44 1280 817304
Fax. +44 1280 817185

Brecknell
100 Armstrong Drive, Fairmont, MN 56031, USA
Email: sales@brecknellsales.com; Website: www.brecknellscales.com
Tel. +1 507 238 4461
Fax. +1 507 238 8271

Bühler Group.
Bühler AG, Gupfenstrasse 5, 9240 Uzwil, Switzerland
Website: www.buhlergroup.com
Tel. +41 71 955 11 11
Fax. +41 71 955 33 79

CN Asia Corporation Bhd.
Chip Ngai Engineering Works Sdn Bhd, Lot 7907, Batu 11, Jalan Balakong, 43300 Seri Kembangan, Selangor, Malaysia
Email: enquiry@cnasia.com; Website: www.cnasia.com
Tel. + 603-8942 6888
Fax. + 603-8942 3365

Clugston Logistics
Clugston Distribution Services, Brigg Road, Scunthorpe,
North Lincolnshire, DN16 1BB, UK
Website: www.clugston.co.uk/logistics
Tel. +44 1724 281281
Fax. +44 1724 270240

County Scales Ltd.
Langley Business Park, Station Road, Nottingham NG16 4DG, UK
Email: sales@countyscales.co.uk; Website: www.countyscales.co.uk
Tel. +44 1773 763555
Fax. +44 1773 763222

Daxner Germany GmbH
Am Breitenstein 3, 97922 Lauda-Königshofen, Germany
Email: office-de@daxner.com; Website: www.daxner.com
Tel. +49 9343 644-163
Fax. +49 9343 644-169

Dingson Food Machinery Ltd.
13 Teng Yun Road, Tanzhou Town, Zhongshan City, Guangdong 528467, China
Email: sales@dsm-mc.com; Website: www.dsm-mc.com
Tel. +86 760 8678 7999
Fax. +86 760 8678 7993

Dingson International Ltd.
Unit A, 7/F Wing On Cheong Building, 5 Wing Lok Street, Central Hong Kong
Email: sales@dsm-intl.com; Website: www.dsm-mc.com
Tel. +852 2545 6122
Fax. +852 2541 9643

Dynamic Air Inc.
1125 Willow Lake Blvd., St. Paul, MN 55110, USA
Email: info@dynamicair.com; Website: www.dynamicair.com
Tel. +1-651-484-2900
Fax. +1-651-484-7015

Farleygreene
Unit 2, Hatch Industrial Park, Greywell Road, Mapledurwell,
Hampshire, RG24 7NG, UK
Website: www.farleygreene.com
Tel. +44 1256 474 547
Fax. +44 1256 329596

Flexicon Corporation
2400 Emrick Blvd., Bethlehem, PA 18020-8006, USA
Email: sales@flexicon.com; Website: www.flexicon.com
Tel. +1 888 353 9426
Tel. +1 610 814 2400
Fax. +1 610 814 0600

Gericke Ltd.
Gerike AG, Althardstrasse 120, 8105 Regensdorf, Switzerland
Email: info@gerike.net; Website: www.gericke.net
Tel. +41 448 71 3636

GKM Siebtechnik GmbH
Felix Wankel Str. 11, DE-74915 Waibstadt, Germany
Email: info@gkm-net.de; Website: https://gkm-net.de
Tel. +49 (7263) 40972-0

Guerin Systems
ZI du Bordage, 49122 Le May-sur-Èvre, France
Website: www.guerinsystems.com
Tel. +33 2 41 63 26 26
Fax. +33 2 4163 1215

Jiangsu Prettech Machinery & Technology Ltd.
A-floor 16, R&D HUB 2, Science and Education Town, No. 801 Changwu Middle Road, Wujin District, Changzhou, Jiangsu, China
Email: info@prettech.cn; Email: admin@prettech.cn; Website: www.sinoprettech.com; Website: www.prettechtank.com
Tel. +86-519-85255001
Fax. +86-519-86713769

Kemutec
130 Wharton Road, Keystone Industrial Park, Bristol, PA 19007, USA
Tel.: +1 215 788 8013
Fax. +1 215 788 5113

Loynds International Ltd.
Units 2-4 Arkwright Court, Blackpool & Fylde Industrial Estate, Blackpool, FY4 5DR, UK
Email: sales@loynds.co.uk; Website: www.loynds.com
Tel. +44 1253 882 961

proBake Inc.
2057 East Aurora Rd., Unit P, Twinsburg, OH 44087, USA
Email: probake@probake.com
Tel. +1 866-812-9654
Fax. +1 330-425-9742

Russell Finex Limited
Russell House, Browells Lane, Feltham, Middlesex TW13 7EW, UK
Email: enquiriesrfl@russellfinex.com; Website: www.russellfinex.com
Tel. +44 207 442 5579
Fax. +44 20 8818 2060

Selis
Eskisehir, Turkey
Email: selis@selis.com.tr
Website: www.selis.com.tr
Tel. +90 222 228 15 66 +90 222 236 12 33
Fax. +90 222 236 11 08 - +90 222 228 15 68

Vac-U-Max
69 William Street, Belleville, NJ 07109, USA
Website: www.vac-u-max.com
Tel. +1 973 759 4600

APPENDIX 2

Manufacturers of Biscuit Production Equipment

Aasted ApS
Bygmarken 7-17, DK 3520, Farum, Denmark
Email:mail@aasted.eu; Website: www.aasted.eu
Tel. +45 44348000

Apex Machinery & Equipment Co. Ltd
Room1203, 12F, 8 YinhuaRoad, Xiangzhou, Zhuhai, Guandong, China
Email:postmaster@apexmeco.com; Website: www.apexmeco.com
Tel. +86 756-2122530/2122531
Fax. +86 756-2122952

Arcall plc
1 Westminster Road, Wareham, Dorset, BH20 4SR, UK
Website:www.tnasolutions.com
Tel. +44 19 295548
Fax. +44 19 295544

Baker Pacific Ltd.
3905 Two Exchange Square, Suite 7443, 8 Connaught Place, Hong Kong
Email:bakerman@bakerpacific.com.hk; Website:www.bakerpacific.com.hk
Tel. +852 2522 1114
Fax. +852 2521 1190

Baker Perkins Ltd.
Manor Drive, Paston Parkway, Peterborough, PE4 7AP, UK
Email:bpltd@bakerperkins.com; Website:www.bakerperkins.com
Tel. +44 1733 283000
Fax. +44 1733 283001

Baker Perkins Inc.
3223 Kraft Ave. S.E., Grand Rapids, MI 49512-2027, USA
Email:bpinc@bakerperkins.com; Website:www.bakerperkins.com
Tel. +1 616 784 3111
Fax. +1 616 784 0973

Bake - o - Nomic Corp.
6 Kiran, K-R-N Society, Off S.V.P. Road, Borivli (West), Mumbai 400 092, India
1, M&n Estate, Opp Thakur Mall, Near Dahisar Check Post, Mumbai,
Maharashtra 401104, India
Email:wonderbake@vsnl.com
Tel. +91 22 2845 0162

Beijing Yinrich Developing Co. Ltd.
No.62 West Street, Gulou, Beijing, China
Website:http://www.yinrich.com
Tel. +86-10-64018055
Fax. +86-10-64073420

Besto Oven Industries
108, Acharya Commercial Center, Dr. C. G. Road, Near Basant Cinema,
Chembur, Mumbai 400074, Maharashtra, India
Website:www.ovenindia.com
Tel. +91-8079452075, +91-22-25218844, +91-22-25568446
Fax. +91-22-25511718, +91-22-25568446

Bühler AG
Gupfenstrasse 5, 9240 Uzwil, Switzerland
Website:www.buhlergroup.com
Tel. +41 71 955 11 11
Fax. +41 71 955 33 79

Cambridge Engineered Solutions
105 Goodwill Road, Cambridge, MD 21613, USA
Email:sales@cambridge-es.com; Website:www.cambridge-es.com
Tel. +1 410 901 2660

Dingson Food Machinery Ltd.
13 Teng Yun Road, Tanzhou Town, Zhongshan City, Guangdong Province, 528467 China
Email:sales@dsm-mc.com; Website:www.dsm-mc.com
Tel. +86 760 8678 7999
Fax. +86 760 8678 7993

Dingson International Ltd
Unit A, 7/F Wing On Cheong Building, 5 Wing Lok Street, Central Hong Kong
Email:sales@dsm-intl.com; Website:www.dsm-mc.com
Tel. +852 2545 6122
Fax. +852 2541 9643

DFE Meincke – Vuurslag B.V.
Schelling Straat 2, 4879 NK, Etten-Leur, Nederland
Email:info@dfemeinckevuurslag.nl; Website:www.vuurslag.nl
Tel. (076) 5720404
Fax (076) 5717077

Dongguang Furong Food Machinery
Nanshe industrial Chashan Town, Dongguan City, Guangdong Province, 523391 China
Email:ftt@dg-furing.cn; Website:www.dg-furong.cn
Tel. +86 769 86171877, 86486035
Fax. +86 769 86402262

Dong Yang Food Machinery Co. Ltd.
301-10 Doksan 1 Dong, Gumchon-gu, Seoul, Korea 153-011
Email:dyfm@chol.com; Website:www.mixer.co.kr
Tel. +82 2 8569759
Fax. +82 2 8530315

Eddingtons Ltd
The Reid Building, New Hayward Farm, Hungerford, Berkshire, RG17 0PZ, UK
Email:sales@eddingtons.co.uk; Website:www.eddingtons.co.uk
Tel. +44 (0)1488 686572

Egan Food Technologies
2001 Waldorf NW, Suite 300, Grand Rapids, MI 49544, USA
Email:info@eganfoodtech.com; Website:www.eganfoodtech.com
Tel. +1 616 855 1820

Esspee Engineers, Kolkata
137, Biplobi Rash Behari Bose Road, 2ndfloor, Kolkata,
West Bengal 700001, India
Diamond Heritage 16, Strand Road, 6th Floor Room No. 627,
Kolkata 700001, India
Website:www.espenger.com
Tel. +91 9 20607 3420

Eurosicma Spa
6 M. Buonarroti St.,20090 Segrate (Mi), Italy
Email:sales@eurosicma.it; Website:www.eurosicma.it
Tel. +39 02 21896.1
Fax +39-02-21896.410

Evergrowing Food Machinery Co. Ltd.
Rm 418, LinHai Building, ShiHua Rd, ZhuHai, China
Website:www.biscuit-machine.net
Tel. +86 756 3321917
Fax. +86 756 3321937

EverSmart
Leilu, ShunDe, Foshan, Guangdong, China
Email:wong@ever-smart.net; Website:www.eversmartpack.com
Tel. +86 757 2220 7105
Fax. +86 757 2220 7106

Fiamma Baking Systems S.r.l.
Via Meucci 8, 37045 Legnago, Italy
Email:2f@2fsrl.net; Website:www.2fsrl.net
Tel. +39 0 442 603 872
Fax. +39 0 442 625 991

Food Processing Equipment s.r.l
Via liariuzzi 15, Loc. San Pancrazio, 43126 Parma, Italy
Email:commerciale@fpe2000.eu; Website:www.fpe2000.it
Tel. +39 0521 303470
Fax. +39 0521 543784

Foshan Shunde Huaji Industrial Co.
China Huaji Industrial Co. Ltd.
Kunzhou Industrial Zone, Beijiao Town, Shunde, Foshan City, Guangdong Province, China
Website:www.sinobake.net
Tel. +86 756 26325603
Fax. +86 756 26325609

Franz Haas
Franz-Haas Strasse, 2100 Lebendorf, Austria
Website:www.haas.com
Tel. +43 2262 600
Fax. +43 2262 600 4040

FRITSCH Bakery Systems GmbH
Bahnhofstraße 27-31, D-97348 Markt Einersheim, Germany
Email:fbs.mail@fritsch-group.com; Website:www.fritsch-group.com
Tel. +49 (0) 93 26/83 - 608
Fax. +49 (0) 93 26/83 - 600

GEA
Beekakker 11, Bakel, 5761 EN, Netherlands
Website:www.gea.com
Tel. +31 492 3493
Fax. +31 492 342985

Haas-Meincke A/S
Tonsbakken 10, 2740 Skovlunde, Denmark
Website:www.haas.com
Tel. +45 77 42 92 00
Fax. +45 77 42 92 01

Hobart Corp.
701 South Ridge Avenue, Troy, OH 45374, USA
Website:www.hobartcorp.com
Tel. +1 888 378 1338

Imaforni Int'l S.p.A.
37030 Colognola ai Colli, Via Strà, 158, Z.I. Verona, Italy
Website:www.imaforni.com; Website:www.gea.com
Tel. +39 045 6174711
Fax. +39 045 6150065

Laser Srl
Via Saturno 36, 37059 Santa Maria di Zevio, Verona, Italy
Website:www.laserbiscuit.it
Tel. +39 0 45 6051428
Fax. +39 0 45 6051426

Machine Builders and Design
806 North Post Road, Shelby, NC 28150, USA
Website:www.machinebuilders.com
Tel. +1 704-482-3456
Fax. +1 704-482-3000

MINIPAN s.r.l.
Via Trebeghino, 47, 48024 Massa Lombarda (Ra), Italy
Email:minipan@minipan.com; Website:www.minipan.com
Tel. +39.0545.971593
Fax. +39.0545.971595

Misuzu BP Corporation
Shiba Fuji Bldg 3F, 1-1-15 Shibadaimon, Minato-Ku, Tokyo, Japan
Website:www.misuzukoki.co.jp
Tel. +81 3343373400

Misuzu Koki Co Ltd (Tokyo office)
2nd Fl Cosmic Building, 1-2-8 Shibadaimon, Minato-Ku, Tokyo, Japan
Website:www.misuzukoki.co.jp
Tel. +81 59 397 3216
Fax. +81 59 397 3217

Morton Mixers and Blenders
Unit 10, Murray Place, Righead Industrial Estate, Bellshill,
Scotland, ML4 3LP, UK
Website:www.morton-mixers.co.uk
Tel. +44 (0)845 2770939
Fax. +44 (0)845 2770949

New Era Machines Pvt. Ltd.
Majara Road, Sahnewal Distt., Ludhiana 141120, India
Email:sales@nem.co.in; Website:www.neweramachines.com
Tel. +91 161 5250299
Fax. +91 161 2510776

Nordic Ware
5005 County Road 25, Minneapolis, MN 55416-2274, USA
Website:www.nordicware.com
Tel. +1 952 920 2888
Fax. +1 952 924 8561

Ovenman Industries Private Ltd
B 14-15 Industrial Area, Site 1, Haripur, Jalalabad, Lucknow Road,
Faizabad 224 001, India
Email:mail@ovenmanindia.com; Website:www.ovenmanindia.com
Tel. +91 983 833000
Fax. +91 5278 236655

Peerless Food Equipment
500 S Vandemark Rd, Sidney, OH 45365, USA
Website:www.peerlessfood.com
Tel. +1 937 492 4158

Ing POLIN & C. SpA
Viale dell'Industria 937135 Verona, Italy
Website:www.polin.it
Tel. +39 045 8289111
Fax. +39 045 8289122

Prater Industries
2 Sammons Court, Bolingbrook, IL 60440, USA
Email:ifo@praterindustries.com; Website:www.praterindustries.com
Tel. +1 630 759 9595
Fax. +1 630 759 6099

Prima Engineering Industries
T -113, Near M.I.D.C. Post Office, Bhosari, Pune 411 026, India
Email:sales@primaengg.com; Website:www.primaengg.com
Tel. +91 20 27120621
Fax. +91 20 27120294

Procys
Parc d'Activités du Pont Béranger, 4 Rue des Frères Lumière,
44680 Saint-Hilaire-de-Chaléons, France
Website:www.procys.eu
Tel. +33 2 3658 2341

Reading Bakery Systems
380 Old West Penn. Avenue, Robesonia, PA 19551, USA
Email:info@readingbakery.com; Website:www.readingbakery.com
Tel. +1 610 6935816
Fax. +1 610 6935512

Rexnord
Corporate Office: 247 Freshwater Way, Suite 200, Milwaukee, WI 53204, USA
Website:www.rexnord.com
Tel. +1 855 663 9876

Rheon Automatic Machinery Co. Ltd.
2-3 Nozawa-machi, Utsunomiya 320-0071, Japan
Email:info@rheon.com; Website:www.rheon.com

Richeng Food Machinery Co. Ltd.
No.2 Dong 6 Road, Jiangyi Leliu, Shunde, Foshan, Guangdong, 528322 China
Email:info@ri-cheng.com.cn; Website:http://www.ri-cheng.com.cn;
Website:www.richeng.en.china.cn
Tel. +86 757 25666837
Fax. +86 757 25668837

Rollmatic
Via Lago di Garda, 112, 36015 Schio (VI), Italy
Email:info@rollmatic.com; Website:www.rollmatic.com
Tel. +39 0445 577000
Fax. +39 0445 579056

RONDO Burgdorf AG
Heimiswilstrasse 42, 3400 Burgdorf, Switzerland
Email:info@rondo-online.com; Website:www.rondo-online.com
Tel. +41 34 420 81 11
Fax +41 34 420 81 99

Schneider Electric SA
43-45, Bd Franklin Roosevelt, 92500 Rueil Malmaison, France
Website:www.schneider-electric.com
Phone: +33(0)1 41 29 70 00
Fax. +33(0)1 41 29 71 00

Spiroflow UK
Lincoln Way, Clitheroe, Lancashire BB7 1QG, UK
sales@spiroflow.com; Website:www.spiroflow.com
Tel. +44 (0) 1200 422525

Spooner Vicars Bakery Systems
Middleby House, Unit 15, Bridge Bank Close, Off Yew Tree Lane, Stonecross Park, Golborne, Wigan, WA3 3JD, United Kingdom
Email:sales@spoonervicarsbakery.com; Website:www.spoonervicarsbakery.com
Tel. +44 151 548 3822
Fax. +44 151 548 2737

Skywin Foodstuff Machinery Co. Ltd.
Joint-venture Industrial Area, West Zone, Yangjiang City, Guangdong, China
Nanchong Industrial Zone, Chencun, Shunde, Foshan City, Guangdong, China
Website:www.skywinbake.com
Tel. +86 757 22237518
Fax. +86 75722237579

tna Solutions Pty Ltd.
24 Carter Street Sydney Olympic Park, NSW 2127, Australia;
PO Box 6007 Silverwater, NSW 2128, Australia
Website:https://www.tnasolutions.com
Tel. +612 9714 2300
Fax. +61 2 9748 2970

TMFCT – Bonnand Lornac
Zone Industrielle, 26 rue de St Rémy, 28270 Brezolles, France
Website:www.bonnandlornac.fr; Website:www.tmfct.fr
Tel. +33 2 37482610
Fax. +33 2 37482582

Werner & Pfleiderer
Industrielle Backtechnik GmbH, Frankfurter Straße 17, 71732 Tamm, Germany
Email:info@wpib.de; Website:www.wpib.de
Tel. +49 7141 202 50
Fax. +49 7141 202 5111

Yesotech
5/46, Jalan TK 3/17, Taman Kinrara, 47190 Puchong, Selangor, Malaysia
Email:yesotech@yahoo.com; Website:www.yesotech.com.my
Tel. +603 8070 1622
Fax. +603 8070 8322

Zhuhai Hong Fu Mechanical Manufacturing Co. Ltd.
China Meihua Road, Xiangzhou Technology Industrial Park,
Zhuhai City, Guangdong Province, 519071 China
Shangchong Science and Technology Industrial District,
Zhuhai, Guangdong Province, China
Website:www.hfoodmachine.cn
Tel. +86 13326631268
Fax. +86 756 8506563

APPENDIX 3

Chocolate Machine Manufacturers

Aasted ApS
Bygmarken 7-17, DK 3520, Farum, Denmark
Email:mail@aasted.eu; Website:www.aasted.eu
Tel. +45 44348000

AMC System Technology (Suzhou) Co. Ltd.
Building 2,No.11 Xuchen Road, SND, Xushuguan District, Suzhou, Jiangsu, China 215151
Email:amc@amcs.cn; Website:www.coolingtunnelsystem.com
Tel. +86 512 68096318
Fax. +86 512 68091880

Apex Machinery & Equipment Co. Ltd.
Rm 1203, 12F, 8 Yinhua Road, Xiangzhou, Zhuhai, China
Email:postmaster@apexmeco.com; Website:www.apexmeco.com
Tel. +86 756 2122530
Fax. +86 756 2122952

Bühler GmbH
Daimlerstrasse 8, 74211 Leingarten, Germany
Email:www.buhlergroup.com; Website:www.hosokawa-bepex.com
Tel. +49 7131 907-0
Fax. +49 7131 907-301

Dedy GmbH
Frankenstraße 152, D-45134 Essen, Germany
Email:info@dedy.de; Website:www.dedy.de
Tel. +49 (0) 201/471031

K&K Kurt Makina
Ege Serbest Bölgesi, Gaziemir, 35410 Izmir, Turkey
Website:www.kurtmakina.com
Tel. +90 - 232 - 251 7299
Fax. +90 - 232 - 251 7275

Lekos, s.r.o.
Trebišov, Kollárova 455, 078 01 Sečovce, Slovak Republic
Email:lekos@lekos.sk; Website:www.lekos.com
Tel. +421 56 6680711
Fax. +421 56 6680713

Mazzetti Renato s.r.l.
Via Monte Grappa, 9, 20067 Tribiano, Milano, Italy
Email:info@mazzettirenato.it; Website:www.mazzettirenato.it
Tel. +39 02 90631159
Fax. +39 02 9064445

Production Techniques Ltd
18 Echelon Place, East Tamaki, Auckland 2013, New Zealand
Email:info@ptl.nz; Website:www.ptl.co.nz
Tel. +64 9 274 3514

Selmi
Via Statale, 151, 12069 Santa Vittoria d'Alba, CN, Italy
Email:info@selmi-group.it; Website:www.selmi-chocolate.it
Tel. +39 0172 479 273/75
Fax. +39 0172 477 814

Sollich KG
Siemensstrasse 17-23, D-32105 Bad Salzuflen, Germany
Email:info@sollich.com; Website:www.sollich.com
Tel.: +49 52 22/95 0 - 0
Fax. +49 52 22/95 0 - 300

APPENDIX 4

Packaging Machine and Equipment Manufacturers

ABB Ltd. (Head Office)
Affolternstrasse 44, CH-8050 Zurich, Switzerland
www.abb.com
Tel. +41 (0)43 317 7111
Fax +41 (0)43 317 4420

ABB Ltd.
Howard Rd, Eaton Socon, Saint Neots, PE19 8EU, UK
Website:www.abb.com; Website:www.new.abb.com/products/robotics
Tel. +44 1480 475321

A.M.P- Rose
Somerby Way, Somerby Park, Gainsborough DN21 1QT, UK
Email:info@amp-rose.com; Website:www.amp-rose.com
Tel. +44 (0) 1427 611969
Fax. +44 (0) 1427 616854

Anton Ohlert GmbH & Co. KG
Goltsteinstrasse 87, 50968 Koeln, Germany
Email:info@ohlert.ru; Website:www.ohlert.com
Tel. + 49 8-10-49 (221) 37 666 - 0
Fax. + 49 8-10-49 (221) 37 666 - 18

Apex Machinery & Equipment Co. Ltd.
Rm 1203, 12F, 8 YinhuaRoad, Xiangzhou, Zhuhai, China
Email:postmaster@apexmeco.com; Website:www.apexmeco.com
Tel. +86-756-2122530/2122531
Fax. +86-756-2122952

ATP Engineering & Packaging S.L
Crta. Barcelona 66, Parc del Molí 10-11 (Entrance on Marconi Street), 08210 Barberà del Vallès, Barcelona, Spain
Website:www.atp-packaging.com
Tel. +34 937 297 980

Avoncourt Packaging Ltd.
Ballycurreen Industrial Estate, Airport Road, Cork, Eire
Email:info@avoncourt.com; Website:www.avoncourt.com
Tel. +353 21 4965691

Robert Bosch GmbH

Packaging Technology
Stuttgarter Straße 130, D-71332 Waiblingen, Germany; P.O. Box 1127, D-71301 Waiblingen
Website:www.boschpackaging.com
Tel. +49 711 811-0
Fax. +49 711 811-45 000

Bradman-Lake Ltd.
Common Lane North, Beccles, Suffolk, NR34 9BP, UK
Email:uk@bradmanlake.com; Website:www.bradmanlake.com
Tel. +44 1502 470 500

Cama Group
Via Como, 9, 23849 Garbagnate Monastero, LC, Italy
Website:www.camagroup.com
Tel. +39 031 879811

Cavanna SpA
Via Matteotti, 104, 28077 Prato Sesia, NO, Italy
Email:cavanna@cavannagroup.com; Website:www.cavanna.com
Tel. +39 163 829111
Fax +39 163 829237

Eurosicma Spa Packaging Solutions
Head Office: 6 M. Buonarroti St., 20090 Segrate, Mi, Italy
Website:www.eurosicma.it

Candy Division
Email:candy@eurosicma.it
Tel. +39 02 21896 401
Fax. +39 02 269529 89

Lines Division
Email:lines@eurosicma.it
Tel. +39 02 21896 403
Fax, +39 0226959 91

Foshan Kendy Intelligent Mechanical Equipment Co., Ltd
Add: No.2 Shizhou, Baichen Road, Gangbei Industrial Park, Chencun Town, Shunde, 528313 Foshan City, Guangdong Province, China
Email:info@kendyautomation.com; Website:www.kendyautomation.com; Website:www.kendyautomation.cn (Alibaba B2B)
Tel. +86 (0)757 8995 5835
Fax. +86 (0)757 2931 3278

Fuji Machinery Ltd.
380 4-Chome, Nakaotai, Nishi-ku, Nagoya 451-568, Japan
Website:www.fuji-machinery.com

Fuji Packaging GmbH
Kattjahren 8, D-22359 Hamburg, Germany
Website:www.fuji-packaging.com
Tel. +49 40 4806110
Fax. +49 40 464063

Hayssen Flexible Systems
225 Spartangreen Blvd., Duncan, SC 29334, USA
Website:www.hayssen.com
Tel. +1 864 4864000

Haysen Flexible Systems
101 Lilac Grove, Beeston, Nottinghamshire, NG9, UK
Website:www.hayssen.com
Tel.: +44 (0) 115 967 8787
Fax. +44 (0) 115 967

Houdijk Holland
Trawlerweg 2-4, 3133 KS Vlaardingen, The Netherlands
Website:www.houdijk.com
Tel. +31 10 4349244
Fax. +31 10 4601335

Ilapack Italia SpA
Via Luciano Lama 11-13, 52045, Foiano della Chiana, AR, Italy
Email:www.ilapack.com; Website:www.ilapak.it
Tel. +39 0575 64421
Fax +39 0575 649753

Innovia Films Ltd.
Wigton, Cumbria, CA7 9BG, UK
Email:filmsinfo@innoviafilms.com; Website:www.innoviafilms.com
Tel. +44 1697 342281

Ishida Co. Ltd
Headquarters, Japan; 44 Sanno-Cho, Shogoin, Sakyo-Ku, Kyoto 606-8392, Japan
Website:www.ishida.com
Tel. +81 75 771 4141
Fax. +81 75 751 1634

Jornen Machinery Co. Ltd.
No.3966 Xinzhuan Road, Songjiang, Shanghai, 201619 China
Email:jornen@jornen.com; Website:www.jornen.com
Tel. +86 21 57678709
Fax. +86 21 57676925

Kawashima Packaging Machinery Ltd.
Tamura Bldg. 16-4 Ginza, 8-Chome, Chuo-ku, Tokyo 104-0061, Japan
Website:www.kawashima-pack.co.jp
Tel. +81 3 5565 6923
Fax. +81 3 5565 5631

Khosla Precisions
93, Industrial Area, Phase – 1, Panchkula - 134108, Haryana, India
Email:Website:www.biscuit-packaging.com
Tel. +91 172 2562664, 2576703

Leeways Packaging Services Ltd
Lobstock, Churcham, Gloucester, GL2 8AN, UK
Website:www.leeways.co.uk
Tel. +44 1452 750487
Fax. +44 1452 750653

Loesch Verpackungstechnik GmbH
Industriestraße 1, 96146 Altendorf, Germany
Email:sales@loeschpack.com; Website:www.loeschpack.com
Tel. +49 9545 449-0

Multihead Weighers
Unit 1 Oakvale House, Burgh Road Industrial Estate, Carlisle, UK
Website:www.multiheadweighers.co.uk
Mobile: +44 77 3065 6691

O-M Ltd.
10-10 Ueno 1-chome, Taito-ku, Tokyo 110-0005, Japan
Email:www.omltd.co.jp; Website:www.ns.0.1.omltd.co.jp
Tel. +81-03-3837-1271

Qingdao Nissin Food Machinery Co. Ltd.
No. 13 Siliu South Road, Qingdao 266042, Shandong, China
Email:export@meetnissin.com; Website:www.meetnissin.com;
Website:www.nissinpacking.en.made-in-china.com
Tel. +86 532 84863994
Fax. +86 532 84863994

Record Packaging Machines
Via Italia 46, Garbagnate Monastero, LC, Italy
Email:info@record.it; Website:www.record.it
Tel. +39 031 850607

Reedbut Group Ltd.
North House 3, Bond Avenue, Bletchley, Milton Keynes, Bucks, MK1 1JJ, UK
Website:www.reedbut.com
Tel. +44 1908 630200
Fax. +44 1908 630210

Riley Product Handling
Unit 2B, Meteor Business Park, Mansfield Road, Derby, DE21 4ST, UK
Email:riley@rileyproducthandling.com; Website:https://www.rileyproducthandling.com
Tel. +44 (0) 1332 866000
Fax. +44 (0) 1332 866127

Schib Packaging Srl
Viale del Lavoro 7/9 A, 36030 Monte di Malo, Vicenza, Italy
Email: info@schib.it; Website: www.schib.it
Tel. +39 0445 602300
Fax. +39 0445 605510

Schur Packaging Systems AB
J.W. Schurs Vej. 1, DK-8700 Horsens, Denmark
Email: sin@schur.com; Website: https://www.schur.com
Tel. +45 7627 2727
Fax. +45 7627 2700

Serpa Packaging Solutions
7020 West Sunnyview Avenue, Visalia, CA 93291, USA
Website: www.serpapackaging.com
Tel. +1 559.651.2339
Fax. +1 559.651.2345

ULMA Packaging
B Garibai, 28, 20560 Oñati, Guipuzkoa, Spain
Website: www.ulmapackaging.com
Tel. +34 943 739200
Fax. +34 943 783218

Universal-Pack
Via Vivare, 425 47842 San Giovanni in Marignano, Rn, Italy
Email: info@universalpack.it; Website: www.universalpack.it
Tel. +39 0541 955276

APPENDIX 5

Quality Control Equipment

Avery Weigh-Tronix LLC
Foundry Lane, Smethwick, West Midlands, B66 2LP, United Kingdom
Email:info@awtxglobal.com; Website:www.averyweigh-tronix.com/uk
Tel. +44 (0)845 307 0314
Fax. +44 (0)870 90 50085

Avery Weigh-Tronix LLC
1000 Armstrong Drive, Fairmont, MN 56031-1439, USA
Email:usinfo@awtxglobal.com; Website:www.averyweigh-tronix.com
Tel. +1 507 238 - 4461
Fax. +1 507 238 - 4195

CCIT
Chemin du Glapin 4, 1162 St-Prex, Switzerland
Website:www.ccit.com
Tel. +41 21 805 0020

CSC Scientific Company, Inc
2799-C Merrilee Drive, Fairfax, VA 22031, USA
Email:csc@cscscientific.com; Website:www.cscscientific.com
Tel. +1 800-621-4778
Fax. +1 703-280-5142

Extech
Nashua, NH 03063, USA
Website:www.extech.com
Tel. +1 877 239 8324 (USA)
Tel. +1 603 324 7800 (Intl.)

FlexPak Leak Detectors Inc.
17769 Centreville Creek Road, Caledon East, Ontario, L7K 2L9, Canada
Website:www.flexpakinc.com
Tel. +1-416-399-5583

ICL Calibration
1501 SE Decker Ave 118, Stuart, FL 34994, USA
Website:www.icllabs.com
Tel. +1 772 286 7710
Fax. +1 772 286 8737

Ishida Co. Ltd.
44 Shogoin Sanno-Cho, Sakyo-Ku, Kyoto 606-8392, Japan
Website:www.ishida.com
Tel. +81 75 771 4141
Fax. +81 75 751 1634

Konica Minolta
JP Tower, 2-7-2 Marunouchi, Chiyoda-ku, Tokyo 100-7015, Japan
Website:www.konicaminolta.com
Tel. +81 3 6250 2111

Metrohm
6555 Pelican Creek Circle, Riverview, FL 33578, USA
Website:www.metrohm.com
Tel. (toll-free): +1 866-METROHM (638-7646)

Metrohm AG
Ionenstrasse, CH 9100 – Herisau, Switzerland
Website:www.chemeurope.com
Tel. +41 71 5110780
Fax. +41 71 3538901

Mettler Toledo
Mettler-Toledo Ltd., 64 Boston Road, Beaumont Leys, Leicester, LE4 1AW, UK
Website:www.mt.com
Tel. 44 116 235 7070

PTI Packaging Technologies and Inspection
145 Main Street, Tuckahoe, New York, 10707, USA
Email:info@ptiusa.com; Website:www.ptiusa.com
Tel. +1 914.337.2005/800.532.1501
Fax. +1 914.337.8519

Rudolph Research Analytical
55 Newburgh Road, Hackettstown, NJ 07840, USA
Email:info@rudolphresearch.com; Website:www.rudolphresearch.com
Tel. +1 973-584-1558
Fax. +1 973-584-5440

Sartorius AG
Weender Landstr. 94-108, 37075 Goettingen, Germany
Website:www.sartorius.com
Tel. +49.551.308.0
Fax. +49.551.308.3289

Unity Scientific
113 Cedar St. Milford, MA 01757, USA
Website:www.unityscientific.com
Tel. 800 445-0759 (USA)
Tel. +1 203 740-2999 (intl.)

Velp Scientifica
Via Stazione 16, 20865 Usmate, MB, Italy
Website:www.velp.com
Tel. +39 039 62 8811
Fax. +39 039 62 88120

APPENDIX 6

Baking Industry Associations

American Bakers Association
601 Pennsylvania Ave., NW, Suite 230, Washington DC 20004, USA
Email:info@americanbakers.org; Website:www.americanbakers.org
Tel. +1 202 7890300

American Institute of Baking (AIB)
AIB International, Inc., PO Box 3999, 1213 Bakers Way,
Manhattan, KS 66505-3999, USA
Website:www.aibonline.org
Tel. +1 785 5374750
Fax. +1 785 537 1493

American Society of Baking
7809 N Chestnut Avenue, Kansas City, MO 64119, USA
Website:www.asbe.org
Tel. +1 800 713 0462
Fax. +1 888 3152612

Bakery Equipment Manufacturers and Allieds (BEMA)
10740 Nall Ave., Suite 230 Overland Park, KS 6621, USA
Email:info@bema.org; Website:www.bema.org
Tel. +1 913 3381300
Fax. +1 913 3381327

Baking Association of Australia
PO Box 574, Gosford, NSW 2250, Australia
Email:tony.smith@baa.asn.au; Website:www.baa.asn.au
Tel. 02 4340 0244
Fax. 02 4340 4288

CAOBISCO Chocolate, Biscuits and Confectionery of Europe
Avenue des Nerviens 9-31, 1040 Brussels, Belgium
Website:www.caobisco.eu
Tel. +32 2 5081021

China Light Industry Machinery Association – CLIMA
RM 3A07, YinGu Plaza, No 9 West Road,
North Fourth Ring Road, Beijing 100080, China
Website:www.clima.net.cn

Cookie and Snack Bakers Association
USA
Craig Parrish: csparrish27@gmail.com; Website:www.casba.us

Federation of Biscuit Manufacturers of India
PHD House, 4/2, Siri Institutional Area, August Kranti Marg, New Delhi 110016, India
Website:www.biscuitfederation.com
Tel. +91-26515137
Fax. +91-26855450

Food and Drink Federation
BCCA Biscuit, Cake, Chocolate and Confectionery Alliance, UK;
Biscuit, Cake, Chocolate and Confectionery Group (BCCC),
6th Floor, 10 Bloomsbury Way, London, WC1A 2SL, UK
Website:www.fdf.org.uk
Tel. +44 20 7836 2460
Fax. +44 20 7836 0580

Japan Food Machinery Association
Fooma Bldg. 3-19-20 Shibaura, Minato-ku, Tokyo 108-0023, Japan
Website:www.fooma.or.jp
Tel. +81-3-5484-0981
Fax. +81-3-5484-0989

Japan Biscuit Association
Website:www.biscuit.or.jp

Japan Bakery and Confectionery Machinery Manufacturers Association
Tatsu Bldg. 5F, 18-2 Ginza 1-chome, Chuo-ku, Tokyo, 104-0061, Japan
Email:mobac@jbcm.or.jp; Website:www.jbcm.or.jp
Tel. +81 3-6673-4333
Fax. +81 3-6673-4568

North American Association of Food Equipment Manufacturers (NAFEM)
161 N. Clark St. Ste. 2020, Chicago, IL 60601, USA
Email:info@nafem.org; Website:www.nafem.org
Tel. +1.312.821.0201

PMMI Association for Packaging and Processing Technologies
11911 Freedom Drive, Ste 600, Reston, VA 20190, USA
Email:info@pmmi.org; Website:www.pmmi.org
Tel. +1 571 61232000
Fax. +1 703 2438556

Process and Packaging Machinery Association (PPMA)
PPMA Ltd, New Progress House, 34 Stafford Road,
Wallington, Surrey, SM6 9AA, United Kingdom
Email:admin@ppma.co.uk; Website:www.ppma.co.uk
Tel. +44 208 773 8111
Fax. +44 208 773 0022

APPENDIX 7

Biscuit Baking Courses

AIB International, Inc.
PO Box 3999, 1213 Bakers Way, Manhattan, KS 66505-3999, USA
Email:info@aibonline.org; Website:www.aibonline.org
Tel. +1 800-633-5137
Tel. +1-785-537-4750
Fax. +1-785-537-1493

Baking Association of Canada
7895 Tranmere Dr., Suite 202, Mississauga, Ontario, L5S 1V9, Canada
Email:info@baking.ca; Website:www.bakingassoccanada.com
Tel. +1 905 4050288
Fax. +1 905 405 0993

Baking Industry Training Australia (BITA)
Unit 5, 54 Riverside Place, Morningside QLD 4170, Australia
Email:nbia@nbia.org.au; Website:www.bita.org.au
Tel. + 61 73899 6100
Fax. + 61 73399 3935

Baking Industry Training College (BITC)
201 Keppel Road, 099419, Singapore
Website:www.bitc.com.sg
Tel. +65 6276 6337
Fax. +65 6276 6608

The Biscuit & Cracker Manufacturers Association (B&CMA)
6325 Woodside Court, Suite 125, Columbia, MD 20723, USA
Website:www.thebcma.org
Tel. +1 443.545.1645
Fax. +1 410.290.8585

Campden BRI
Station Road, Chipping Campden, Gloucestershire UK. GL55 6LD, UK
Email:information@campdenbri.co.uk; Website:www.campdenbri.co.uk
Tel. +44(0)1386 842000
Fax. +44(0)1386 842100

Cornwall College Group, UK
Email:enquiries@cornwall.ac.uk; Website:www.cornwall.ac.uk
Tel. +44 330 1232523

IFT Institute of Food Technologists
525 W.Van Buren, Ste 1000 Chicago, IL 60607, USA
Email:info@ift.org; Website:www.ift.org
Tel. +1.312.782.8424
Fax. +1.312.782.8348

International School of Baking (ISB)
1971 NW Juniper Street, Bend, OR 97703, USA
Website:www.schoolofbaking.com
Tel. +1 541-389-8553

IPB – Institut Pertanian Bogor
Campus IPB Darmaga, Bogor 16680, West Java, Indonesia
Email:ico@ipb.ac.id; Website:www.ipb.ac.id Website:www.ico.ipb.ac.id
Tel. +62 251 8622638/ 8627916
Fax. +62 251 8622638/ 8627916

King Arthur Flour
Skagit Valley, Washington State, USA; Norwich, Vermont, USA
Email:bakingeducation@kingarthurflour.com; Website:www.kingarthurflour.com/baking-school
Tel. +1 8 00 652 3334

National Bakery School, London
103 Borough Road, London SE1 0AA, UK
Website:www.lsbu.ac.uk
Tel. +44 20 7815 6189 (International)
Tel. 0800 923 8888 (General), 020 7815 7815 (Courses)

US Wheat Associates
3103 10th Street North, Suite 300, Arlington, VA 22201, USA
Email:infoARL@uswheat.org; Website:www.uswheat.org/customers
Tel. +1 202 4630999
Fax. +1 703 5244399

UTC – University Technical Colleges
(for 14-19 year old students), Baker Dearing Educational Trust,
4 Millbank, London SW1P 3JA, UK
Email:director@utcolleges.org; Website:www.utcolleges.org
Tel. +44 20 7960 1555

INDEX

Note: Page numbers followed by *f* indicate figures and *t* indicate tables.

FSC
www.fsc.org
MIX
Papier | Fördert gute Waldnutzung
FSC® C083411

Kolibri 360 GmbH
Ferdinand-Jühlke-Straße 7
99095 Erfurt, Deutschland
produktsicherheit@kolibri360.de